ISBN 978-3-540-05102-2   ISBN 978-3-540-36330-9
DOI 10.1007/978-3-540-36330-9

15. Band, 4. Heft

Fortschritte der chemischen Forschung
Topics in Current Chemistry

Herausgeber:

Prof. Dr. *A. Davison*	Department of Chemistry, Massachusetts Institute of Technology, Cambridge, MA 02139, USA
Prof. Dr. *M. J. S. Dewar*	Department of Chemistry, The University of Texas Austin, TX 78712, USA
Prof. Dr. *K. Hafner*	Institut für Organische Chemie der TH 6100 Darmstadt, Schloßgartenstraße 2
Prof. Dr. *E. Heilbronner*	Physikalisch-Chemisches Institut der Universität CH-4000 Basel, Klingelbergstraße 80
Prof. Dr. *U. Hofmann*	Institut für Anorganische Chemie der Universität 6900 Heidelberg 1, Tiergartenstraße
Prof. Dr. *K. Niedenzu*	University of Kentucky, College of Arts and Sciences Department of Chemistry, Lexington, KY 40506, USA
Prof. Dr. *Kl. Schäfer*	Institut für Physikalische Chemie der Universität 6900 Heidelberg 1, Tiergartenstraße
Prof. Dr. *G. Wittig*	Institut für Organische Chemie der Universität 6900 Heidelberg 1, Tiergartenstraße

Schriftleitung:

Dipl.-Chem. *F. Boschke*	Springer-Verlag, 6900 Heidelberg 1, Postfach 1780

Springer-Verlag	6900 Heidelberg 1 · Postfach 1780 Telefon (06221) 49101 · Telex 04-61723 1000 Berlin 33 · Heidelberger Platz 3 Telefon (0311) 822001 · Telex 01-83319
Springer-Verlag New York Inc.	New York, NY 10010 · 175, Fifth Avenue Telefon 673-2660

Introduction to

All-Valence Electrons S.C.F. Calculations of Large Organic Molecules

Theory and Applications

Gilles Klopman and Brian O'Leary

Case Western Reserve University, Cleveland, Ohio 44106, USA

Contents

445

I. Introduction

The concept of molecular orbitals being constructed from atomic orbitals was suggested as early as 1929 by Lennard-Jones [1] and subsequently referred to by Mulliken [2] as the *"linear combination of atomic orbitals"* (L.C.A.O.-M.O.) approach. It met its first success when Hückel [3], in 1931, took the idea as the basis of his molecular orbital theory for conjugated hydrocarbons. In these compounds, usually planar, each carbon atom is surrounded by only three neighbors. Such a conformation can best be explained by allowing the carbon atomic orbitals to hybridize into three sp^2 coplanar orbitals at 120° from each other and one p_π orbital perpendicular to this plane. Hückel noticed, that in such an arrangement, the p_π orbitals of the various carbon atoms overlap only between themselves and not with the sp^2 orbitals. Hence sigma-pi separation was achieved and a good description of the π systems could be obtained by a linear combination of the p_π atomic orbitals, providing however that electron repulsions with the σ framework are neglected (or averaged).

The success of such a description, however, rests upon the assumption that the chemical behavior of conjugated hydrocarbons is solely determined by its π electrons. Its advantage is that only one orbital per carbon atom is involved in the L.C.A.O. calculation, and the hydrogen ls orbitals can be neglected altogether.

The remaining difficulties in solving the L.C.A.O. problem were left by rather trivial assumptions such as the neglect of electron repulsion (one-electron theory), of overlap and of long-range (non-bonding) interactions.

In spite of these gross approximations, the method proved to be extremely useful and was extensively used to correlate the chemical properties of conjugated systems. Several attempts were subsequently made to introduce the repulsions between the π electrons in the calculations. These include the work of Goeppert-Mayer and Sklar [4] on benzene and that of Wheland and Mann [5] and of Streitwieser [6] with the ω technique. But the first general methods of wide application were developed only in 1953 by Pariser and Parr [7] (interaction of configuration) and by Pople [8] (SCF) following the publication by Roothaan [9] of his self-consistent field formalism for solving the Hartree-Fock equation for molecules. All these methods, however, still make use of the $\sigma\pi$-separability and only the one p_π orbital per carbon atom is included specifically in the calculations.

The *all-valence orbital methods*, on the other hand, are methods using as a basis set all atomic orbitals having the same quantum number as the highest occupied orbital. Such methods represent a comparatively recent development, and have their origin in the one-electron theories applied to calculations on paraffins [10]. The rather slow development in this field (the first methods were suggested only in the 50's) may be attributed, firstly, to the overshadowing success of the π calculations; secondly, and perhaps more important, to the fact that in saturated hydrocarbons a number of properties are given fairly accurately by simply adding the properties of the isolated bonds, and those that are not were either not then known, e.g. spectra, or very small, e.g. dipole moments. The one exception is *ionization potential*, which is not a regular function of each bond. For these reasons the early theories were concerned in the main with specific properties of saturated hydrocarbons. Thus we have Brown's [11] linear combination of bond orbitals (L.C.B.O.) theory and the perturbation work of Dewar and Pettit [12] concerned with bond energies and heats of formation, while the equivalent orbital theory of Lennard-Jones and Hall [13] and the united atom theory of Franklin [14] were concerned with ionization potentials.

In 1954 Sandorfy and Daudel [15] published their *"C" approximation*, a one-electron approximation which employs a linear combination of sp^3 orbitals and borrows most of its simplifying features from the Hückel theory. The originality of this method lies essentially in the introduction of a resonance integral $m\beta$ between sp^3 orbitals of the same carbon atom. Sandorfy [16] showed that the inductive effect due to a heteroatom can be reproduced by such a calculation.

Yoshimuzi [17], using this method with different values for m, calculated the electronic distribution produced by substituting a heteroatom for hydrogen. He found that a value of the m parameter such that $m^2 = 0.12$ was necessary in order to reproduce the dipole moments of a set of linear paraffins. Fukui et al. [18], using the positive square root of 0.12, i.e. $m = +0.35$, were able to correlate the ionization potentials, heats of formation, and bond energies in linear as well as cyclic hydrocarbons and their derivatives. It was also shown that the method permits a coherent interpretation of inductive effects to be made so that a relation exists between some calculated values and the reactivity.

There was, however, one serious shortcoming. The method did not work well for *branched hydrocarbons*. To overcome this, Fukui introduced further parameters in order to give different α values (diagonal matrix element) to primary, secondary, tertiary, and quaternary carbon centers.

Klopman [19] pointed out that the method might be placed on a more theoretically satisfying basis by replacing the interaction between orbitals of the same atom by the interaction between non-bonded orbitals. This was amplified by the fact that if the hydrogen atoms are specifically included, the method yields simple additivity of bond properties, thus losing its usefulness. He next demonstrated that by taking m as the negative square root of 0.12, i.e. $m = -0.35$, one could produce results equivalent in accuracy to Fukui's without the necessity of special parametrization for the branched-chain case.

Side by side with these developments in the organic field, Wolfsberg and Helmholz [20] published in 1952 one of the earliest one-electron all-valence orbital calculations on the electronic structure and spectra of the *inorganic oxyanions* MnO_4^-, CrO_4^{2-} and ClO_4^-. Their method requires the evaluation of two integrals, the diagonal matrix element H_{ii} and the off-diagonal element H_{ij}. They assumed, first, that non-valence electrons were unaffected by bonding and could thus be considered as constituting undisturbed cores; secondly, the value of H_{ii} could be equated, to a first approximation, to the valence state ionization potential; third, that H_{ij} could be evaluated by means of the expression

$$H_{ij} = \frac{KS_{ij}(H_{ii} + H_{jj})}{2} \qquad (1)$$

originally due to Mulliken [21].

The result of these efforts, in both the organic and inorganic fields, was the gradual build-up of a set of rules to be used for the approximations in a one-electron all-valence orbital treatment of molecules. This culminated in Hoffmann's *"extended Hückel theory"* where these and a few new rules were brought together in a very rational manner leading finally to a coherent one-electron method of wide application [22].

Within the last five years the development of large-capacity computers has been paralleled by the development of methods for performing all-valence electron calculations including electron interaction for large molecules. In 1964 two papers published independently by Pohl [23] and by Klopman [24] laid out the procedure for such calculations; however, they were restricted to small molecules. The *self-consistent field* (S.C.F.) formalism which is used in these methods was found to be particularly versatile and appropriate for computer use.

Because of the electron interaction terms, the matrix elements in this approximation are themselves a function of the coefficients of the atomic orbitals. Initially, therefore, a reasonable guess is made as to the electronic distribution in the molecule and the calculation carried out until a new charge distribution is obtained. This new set is then used as a starting point and the procedure repeated until self-consistency is attained. Hence, the final charge distribution is independent of the initial choice and the total energy obtained for this self-consistent charge distribution is the energy of the molecule. An excellent discussion of the various means by which a coherent S.C.F. theory for all valence electrons could be attained was published by Pople, Santry and Segal [25] in 1965. In this article, the principles of several methods of increasing levels of sophistication were outlined. Among these, the CNDO, complete neglect of differential overlap, was the first to be developed, and applied to the calculation of charge densities in large organic molecules [26]. This was followed in 1967 by Pople's *intermediate neglect of differential overlap* (INDO) [27] and by Dewar and Klopman's *partial neglect of differential overlap* (PNDO) [28] approach. Since then, a number of variants, usually differing from the preceding ones only by the choice of parameters or the method or parametrization have been published [29], and were first reviewed by Jaffe [29d] in 1969. These methods, as well as the first ones, are discussed in the following pages. Their objective is usually to permit a more accurate calculation of a specific property and are parametrized accordingly. No program based on the NDDO method (*neglect of diatomic differential overlap*), which is the most sophisticated method yet suggested, has been reported so far [a].

---

[a] While this review was in press, such a method was published by Sustmann, Williams, Dewar, Allen and Von Schleyer [58].

II. The L.C.A.O.M.O. Procedure, Hartree-Fock and S.C.F. Formalism

The Hartree-Fock method is a procedure for finding the best *"many-electron" wave function* Ψ as an antisymmetrized product of *one-electron orbitals* $\phi_m(\mu)$

$$\Psi = \frac{1}{\sqrt{N}} \sum_p (-1)^P P[\phi_1(1)\,\alpha(1)\,\phi_2(2)\,\beta(2)\ldots\phi_m(\mu)\,\sigma(\mu)\ldots\phi_n(n)\,\sigma(n)] \quad (2)$$

In the case of molecules, the functions $\phi_m(\mu)$ are molecular orbitals formed usually from a *linear combination of atomic orbitals* χ_i (L.C.A.O.M. O. approximation)

$$\phi_m(\mu) = \sum_i a_{mi}\,\chi_i(\mu) \quad (3)$$

The set of *initial atomic functions* χ_i is called the *basis set*. Although the complete solution of the Hartree-Fock problem requires an infinite basis set, good approximations can be achieved with a limited number of atomic orbitals. The minimum number of such functions corresponding approximately to the number of electrons involved in the molecule is the *"minimal" basis set*. The coefficients a_{mi} which measure the importance of each atomic orbital in the respective molecular orbitals are parameters determined by a variational procedure, i.e. chosen so as to minimize the expression

$$E = \frac{\int \Psi H \Psi\, dt}{\int \Psi \Psi\, dt} \quad (4)$$

where E represents the *expectation value* of the electronic energy associated with the Hamiltonian H of the given molecule. This *Hamiltonian* is espressed as

$$H = \sum_\mu \left(-\frac{1}{2}\nabla_\mu^2 - \sum_A \frac{Z_A}{r_{A\mu}}\right) + \sum_{\mu > \nu}\sum \frac{1}{r_{\mu\nu}} \quad (5)$$

451

where $-1/2\nabla_\mu^2$ represents the kinetic energy operators of the individual electrons μ,

$-\sum \dfrac{Z_A}{r_{A\mu}}$ are the potential energy operators

where Z_A is the charge on nucleus A, and

$r_{A\mu}$ is the distance between this nucleus and electron μ

$1/r_{\mu\nu}$ are the mutual repulsion operator between any two electrons μ and ν.

Representing the parenthesized one-electron part of the Hamiltonian by $H(\mu)$, we may rewrite the above equation as

$$H = \sum_\mu H(\mu) + \sum_{\mu\,>\,\nu}\sum \frac{1}{r_{\mu\nu}} \tag{6}$$

The variation theorem requires, for each molecular orbital m, that the coefficients a_{mi} satisfy the following sets of simultaneous equations:

$$\sum_i a_{mi}\,(F_{ij} - E_m S_{ij}) = 0 \qquad \text{for } j = 1, 2, \ldots, N \tag{7}$$

where N is the *number of basis set functions* used,

and $\sum_i \sum_j a_{mi}\, a_{mj}\, S_{ij} = 1$ (the normalization conditions)   (8)

where S_{ij} is the *overlap*, equal to $\int \chi_i \chi_j\, dt$.

The solution of the secular equation

$$|F_{ij} - E S_{ij}| = 0 \tag{9}$$

are the values E_m which satisfy the first set of simultaneous equations (Eq. 7).

Roothaan has shown that for a closed shell system, F_{ij} is given by

$$F_{ij} = H_{ij}^c + \sum_k \sum_l P_{kl}\,[(ij\,|\,kl) - \tfrac{1}{2}(ik\,|\,jl)] \tag{10}$$

where H_{ij}^c is given by

$$H_{ij}^c = \int \chi_i\,(\mu) H(\mu) \chi_j(\mu)\, dt_\mu \tag{11}$$

P_{kl} is the *total electronic population* in the overlap region between atomic orbitals k and l:

$$P_{kl} = 2 \sum_{m}^{occ} a_{mk} \, a_{ml} \tag{12}$$

and

$$(ij\,|\,kl) = \int\int \chi_i(\mu)\, \chi_k(\nu)\, \frac{1}{r_{\mu\nu}}\, \chi_j(\mu)\, \chi_l(\nu)\, dt_\mu\, dt_\nu \tag{13}$$

The solution of the secular equation $|F_{ij} - ES_{ij}| = 0$ requires the evaluation of the *constituent matrix terms* F_{ij}. The F_{ij}'s are, however, themselves functions of the coefficients of the atomic orbitals a_{mi} through P_{kl} and therefore can only be evaluated by solving the secular equation. The Hartree-Fock procedure thus requires that a *preliminary guess* be made as to the values of the molecular population distribution terms P_{kl}; these values are then used to calculate the matrix elements F_{ij} and thence solve the secular determinant. This, in turn, provides a better approximation to the wave function and an „improved" set of values of P_{kl}. The above procedure is repeated with this first improved set and a second improved set evaluated. The process is repeated until no difference is found between successive improved wave functions. Finally, it may be shown that when such a calculation has been iterated to self-consistency the *total electronic energy E of a closed shell molecule* is given by

$$E = \sum_i \sum_j P_{ij} \left[H_{ij}^c + \tfrac{1}{2} \sum_k \sum_l P_{kl} \left(<ij\,|\,kl> - \tfrac{1}{2} <ik\,|\,jl> \right) \right] \tag{14}$$

The main obstacles to the solution of this problem lie in the formidable number of multicentered integrals $<ij\,|\,kl>$ which arise even with the use of a minimal basis set, and the difficulty involved in their evaluation. This is illustrated in Table 1, where the number of electron interaction integrals is computed for a minimal basis set calculation of various compounds. The total number of such bielectronic integrals can be computed by the following equation.

$$\frac{n}{4}(n+1)\left[\frac{n}{2}(n+1)+1\right] \tag{15}$$

where n is the number of functions of the minimal basis sets. In the case of the *hydrogen molecule*, the number of such electron interaction integrals amounts to six. If, however, the basis set is extended to include the L shell as well, then 1540 electron-electron interaction integrals must be evaluated. An even more dramatic increase would be observed for the other compounds even if the basis set was only slightly extended.

Table 1. *Total number of bielectronic integrals resulting from a minimal basis set. (These numbers have not been corrected to take into account the fact that some integrals are identical due to the symmetry of the molecules.)*

Integrals	Compounds			
	H_2	CH_4	C_2H_6	C_3H_8
Basis set	2	9	16	23
1-center	2	124	246	368
2-center	4	464	2680	6652
3- and 4-center	—	447	6390	31,206
Total	6	1035	9316	38,226

Even when the Hartree-Fock solution can be attained, we are still nowhere near the solution of the Schrödinger equation since, due to the original choice of the wave function as being a product of one-electron orbitals, a far more fundamental difficulty arises. In the "product of one-electron orbitals" approximation, the probability of finding an electron at a certain point in space is not affected by the fact that another electron might already occupy that position. An immediate consequence of the neglect of "electron correlation" is that the calculated electron repulsion energies will be found to be larger than would be the case if the tendency for the electrons to avoid each other were properly taken into account. Unfortunately this difficulty is implicit in the Hartree-Fock procedure and cannot be overcome unless correction terms involving the coordinate of more than one electron simultaneously are incorporated.

Despite these shortcomings, the Roothaan equation has been used extensively and the Hartree-Fock energies of various small molecules have been calculated. However, the difficulties encountered in calculating the energy of large molecules are such that *simplified methods* are desirable in these cases. Several such methods will be discussed in the next section;

their objective is to reduce the size of the problem without losing too much of the significance of the results. This is done by neglecting the largest possible number of "hopefully" less important integrals and evaluating those remaining either by simple semi-empirical methods or even by direct comparison with experimental data. This allows most of the correlation energy and part of the error introduced by neglecting numerous integrals to be averaged and usually accounted for.

III. The All-Valence Electrons, Neglect of Diatomic Differential Overlap Method

A. General Approximations

Among the numerous approximations which could be used to simplify the Hartree-Fock problem, the all-valence electrons, N.D.D.O. method is particularly appropriate, due to the simplicity and adequacy of its approximations. These are:

1. Only valence electrons are accounted for specifically.

2. Only atomic orbitals of the same principal quantum number as that of the highest occupied orbital in the isolated atom are included in the basis set.

3. Diatomic differential overlap is neglected, i.e. $\chi_i (\mu) \chi_j (\mu) = 0$ if the orbitals χ_i and χ_j are not on the same atoms. Hence the overlap is neglected,

$$S_{ij} = \int \chi_i (\mu) \chi_j (\mu) = 0 \tag{16}$$

and

$$\iint \chi_i (\mu) \chi_k (\nu) \frac{1}{r_{\mu\nu}} \chi_j (\mu) \chi_l (\nu) dt_\mu dt_\nu = (ij\,|\,kl) = 0 \tag{17}$$

unless χ_i and χ_j are atomic orbitals belonging to the same atom A and χ_k and χ_l are atomic orbitals belonging to the same atom A, or B.

The first of these approximations *allows us to neglect the inner electrons of the atom* by treating them as part of a core whose charge will be approximately equal to that of the nucleus minus one per core electron. The second approximation considerably *reduces the initial number of integrals*. At first sight, it might be thought that the inclusion of more orbitals in the basis set would automatically improve the results. The benefit gained by adopting such a plan is, however, made negligible by the overriding effects of the neglect of correlation energy and of the other approximations. *The third approximation removes all three- and four-center integrals* and some two-center integrals. With these approximations, the matrix elements in the secular equation become:

$$F_{ij} = H_{ij}^c + \sum_B \sum_{k,l} {}^B P_{kl} \, (ij \,|\, kl) - \tfrac{1}{2} \sum_{k,l} {}^A P_{kl} \, (ik \,|\, jl) \quad (i, j \text{ both on atom A})$$

$$F_{ij} = H_{ij} - \tfrac{1}{2} \sum_k {}^A \sum_l {}^B P_{kl} \, (ik \,|\, jl) \quad (i \text{ on atom A, } j \text{ on atom B})$$

Table 2 illustrates the overall effect of both these approximations, and other ones to be dealt with subsequently, on the number of one-, two-, three- and four-center integrals involved in a calculation on propane.

Table 2. *Number of bielectronic integrals involved in the calculation of propane by various approximations*

Integrals	H-F Minimal basis set	NDDO	PNDO	(M)INDO	CNDO
1-center	368	173	14	26	11
2-center	6652	568	307	55	55
3—4 center	31,206	0	0	0	0
Total	38,226	741	321	81	66

Even with the simplifications we have outlined, there still remains the problem of too many integrals. Further simplification, however, becomes critical to the method itself due to invariance requirements.

B. The Invariance Requirements

Pople *et al.* [25] pointed out that while the results obtained for two-center integral evaluation in a full Roothaan S.C.F.M.O. treatment are independent of the choice of axis, the same is not true in simplified versions. Such integrals are sensitive to the choice of coordinate system and the hybridization of the orbitals.

Accordingly, the results of the simplified versions are required to be invariant to two types of transformation:

1. A unitary transformation between the various orbitals of an individual atom; we shall refer to this as *space invariance*.

2. A unitary transformation between the s and p orbitals of an individual atom; we shall refer to this as *hybridization invariance*.

A simple physical picture will serve to illustrate the significance of invariance.

1. Space Invariance

It is well known that n degenerate solutions of the Schrödinger equation belonging to a particular energy level can be made to span an n-dimensional function space. Thus every solution of the equation with this energy can be expressed in terms of n linearly independent wave functions $\chi_1 \chi_2 \ldots \chi_n$. The level being referred to is n-fold degenerate. Although an s state is non-degenerate, a p state is three-fold degenerate and the resulting function space is conveniently spanned by the three mutually orthogonal p_x, p_y and p_z orbitals. Thus although a specific p orbital may be defined as one of the basis functions, p_x say, it can equally well be expressed as a linear combination of $p_{x'}$ and $p_{y'}$, $1/\sqrt{2}\,(p_{x'}+p_{y'})$, say, where the atomic orbital basis set has undergone a simple unitary transformation (a rotation about p_z) and the basis functions $p_{x'}$ and $p_{y'}$ happen to be now at 45° to p. Consider the valence s atomic orbital of an atom A and a valence p atomic orbital of an atom B in the molecular entity:

The *two-center two-electron repulsion integral* $(ii|jj)$ can be expressed as

$$(ii|jj) = (ss|xx) \quad \text{for } p = p_x \tag{18}$$

or

$$(ii|jj) = \frac{1}{2}\{(ss|x'x') + 2(ss|x'y') + (ss|y'y')\} \quad \text{for } p = \frac{1}{\sqrt{2}}(p_{x'}+p_{y'}) \tag{19}$$

Since the energy of the electron repulsion integral must in each case be the same, then

$$(ss|xx) = \tfrac{1}{2}\{(ss|x'x') + 2(ss|x'y') + (ss|y'y')\} \tag{20}$$

There are two ways of achieving the above equality. We may assume that the electron repulsion is completely independent of the orientation of the orbitals, hence,

$$(ss|xx) = (ss|x'x') = (ss|y'y') \tag{21}$$

and set $(ss|x'y') = 0$ in order to maintain the equivalence between the two schemes. This amounts to treating the valence orbitals on B as spherically symmetrical.

Alternatively we can say that the electron repulsion term does depend on orbital orientation, i.e.,

$$(ss\,|\,xx) \neq (ss\,|\,x'x') = (ss\,|\,y'y') \tag{22}$$

and therefore $(ss\,|\,x'y')$ must be different from zero in order to maintain the above equality (Eq. 20).

The above argument must also apply to the two-center one-electron integrals H_{ij}. However, here the problem can be corrected easily by setting these integrals proportional ot the overlap integral.

2. Hybridization Invariance

Except for the fact that we are now dealing with hybrid s and p orbitals on atom B, the problem is completely analogous to that of space invariance. Hybridization invariance is of less importance than space invariance providing a rigorous selection of standards is made.

C. The Methods

The requirements of space invariance restrict further simplification of the Hartree-Fock problem to one of two distinct routes.

1. CNDO and (M)INDO

The assumption

$$(ss\,|\,xx) = (ss\,|\,x'x') = (ss\,|\,y'y') \text{ and } (ss\,|\,x'y') = 0 \tag{23}$$

requires that all two-center integrals involving the differential overlap between two orbitals on the same atom be neglected. Pople *et al.*, as well as adopting this in their first all-valence electrons complete neglect of differential overlap (CNDO) method, went one step further in neglecting one-center electron interactions involving differential overlap [26]. With these approximations the Hartree-Fock matrix elements become:

$$F_{ii} = u_{ii} + (P_{AA} - \tfrac{1}{2} P_{ii})\, \Gamma_{AA} + \sum_{B \neq A} (P_{BB}\, \Gamma_{AB} - v_{AB})$$

$$\tag{24}$$

$$F_{ij} = H_{ij} - \tfrac{1}{2} P_{ij}\, \Gamma_{AB} \quad (i \neq j)$$

where the atomic orbital ψ_i is centered on atom A and ψ_j on atom B.

In deriving these equations Pople separated the core matrix element H_{ii} thus:

$$H_{ii} = (ii \,|\, -\tfrac{1}{2}\nabla^2 - v_A \,|\, i) - \sum_{B \neq A} (i \,|\, v_B \,|\, i) \tag{25}$$

$$= -u_{ii} - \sum_{B \neq A} v_{AB} \tag{26}$$

Here u_{ii} is that part of the diagonal matrix element involving the one-electron Hamiltonian containing only the core of its own atom and v_{AB} gives the interaction of an electron in ψ_i on atom A with the cores of other atoms B.

The *electron interaction integrals* $(ii \,|\, jj)$ are written Γ_{AB} and are assumed to depend only on the atoms A or B to which ψ_i and ψ_j belong and not on the type of orbital. P_{ij} are the components of the charge density and bond order matrix

$$P_{ij} = 2 \sum_{m}^{occ} a_{mi}\, a_{mj} \tag{27}$$

P_{AA} is the *total charge density on atom a*

$$P_{AA} = \sum_{i}^{A} P_{ii} \tag{28}$$

Finally it may be shown that under the CNDO approximation, the total energy of a molecule can be expressed as a sum of one- and two-atom terms

$$E_{total} = \sum_{A} E_A + \sum_{A<B} E_{AB} \tag{29}$$

where

$$E_A = \sum_{i}^{A} P_{ii}\, u_{ii} + \tfrac{1}{2} \sum_{i}^{A} \sum_{j}^{A} (P_{ii}\, P_{jj} - \tfrac{1}{2} P_{ij}^2)\, \Gamma_{AA} \tag{30}$$

and

$$E_{AB} = \sum_{i}^{A} \sum_{j}^{B} (2\, P_{ij}\, H_{ij} - \tfrac{1}{2} P_{ij}^2\, \Gamma_{AB}) + (Z_A Z_B R_{AB}^{-1}$$
$$- P_{AA}\, v_{AB} - P_{BB}\, v_{BA} + P_{AA}\, P_{BB}\, \Gamma_{AB}) \tag{31}$$

where R_{AB} is the distance between atoms A and B.

The neglect of the one-center electron interactions involving differential overlap between two orbitals results in certain one-center exchange integrals such as $(2s\,2p_x|2s\,2p_x)$ being omitted. This renders the method incapable of introducing quantitatively *Hund's rule effects*, i.e. that two electrons on different atomic orbitals on the same atom have a lower repulsion energy if their spins are parallel. Although this makes the scheme too restricted for molecular spectroscopy (it cannot resolve degeneracy), the omission is not too serious provided calculation is restricted to the ground states of closed-shell molecules.

The scheme does have the advantages of simplicity and can be carried over even for large molecules since there is only one one-center and one two-center bielectronic integrals per pair of atoms, irrespective of the number of functions used for each of them.

Several schemes based on the CNDO approximation have been proposed [30-32]. Some differ only in the choice of approximation used to calculate semiempirically the various remaining integrals. Others introduce additional features which make this method particularly suitable for certain purposes. Thus Jaffe and Del Bene [31] developed a modified version of the CNDO procedure which includes some configuration interaction and makes it extremely useful for calculating *spectroscopic terms*. The exchange modified zero differential overlap (EMZDO) method proposed by Dixon [33] and the almost identical intermediate neglect of differential overlap (INDO) method introduced by Pople et al. [27] and the modified intermediate neglect of differential overlap (MINDO) method introduced by Dewar et al [34,35] all retain only the main requirement of *rotational invariance*, i.e. that only the two-center two-electron interaction involving differential overlap between orbitals on the same atom need be neglected. This means that certain exchange integrals of the form $(2s\,2p_x|2s\,2p_x)$ are retained and thus the qualitative effects of Hund's rule may be introduced, making the method particularly appropriate for open-shell systems. INDO was developed with special emphasis on the calculation of free radical properties such as *E.S.R. spectra*. Pople pointed out that for an open-shell system a restricted wave function (i.e. each orbital ψ_i is doubly occupied, thus restricting the α and β electrons to identical spatial orbitals) prevents a realistic description of the unpaired spin distribution in a system. He therefore developed INDO in terms of an unrestricted molecular wave function (i.e. one in which different spatial orbitals describe the motion of electrons with different spins).

The unrestricted L.C.A.O.–S.C.F. method reduces to the restricted method when α and β electrons are assigned to spatially identical molecular orbitals. Thus under the INDO method the Hartree-Fock matrix elements for an open-shell system become

$$F_{ii}^a = u_{ii} + \sum_{kl}^{A} \{P_{kl}\,(ii\,|\,kl) - P_{kl}^a\,(ik\,|\,il)\}$$

$$+ \sum_{B \neq A} (P_{BB}\,\Gamma_{AB} - v_{AB}) \qquad (i \text{ on atom A}) \qquad (32)$$

$$F_{ij}^a = u_{ij} + \sum_{kl}^{A} \{P_{kl}\,(ij\,|\,kl) - P_{kl}^a\,(ik\,|\,jl)\} \quad (i \neq j \text{ both on atom A})$$

$$F_{ij}^a = H_{ij} - P_{ij}^a\,\Gamma_{AB}$$

Here the notation is as used in CNDO. The F_{ij}^β elements for electrons of opposite spin have the same form.

The results of INDO are apparently very similar to those of CNDO when the same set of approximations are used to calculate common integrals. Indeed, the choice of approximations to be used in the (M)INDO method are, as in the other schemes, dictated by the objectives of the method (or the authors' preference). The MINDO method is especially parameterized to calculate *heats of formation* and MINDO/2 [35], which has been specially reparameterized, claims both *heats of formation* and *bond distances*. Both of these methods are also suitable for the calculation of open shell systems. An additional approximation, however, was made in order to achieve this, namely that an electron can be treated as half an electron pair (i. e. two halves of one electron).

2. PNDO

The partial neglect of differential overlap (PNDO) method, originally referred to as the partial neglect of diatomic differential overlap (PNDDO) introduced by Dewar and Klopman [28], is to date the only method to adopt

$$(ss\,|\,xx) \neq (ss\,|\,x'x') = (ss\,|\,y'y') \quad \text{and}$$

$$(ss\,|\,x'y') \neq 0$$

the alternative route dictated by rotational invariance considerations. The PNDO method retains only those integrals necessary for the maintenance of rotational invariance. Thus since the two-center $(ij\,|\,kl)$ integrals involving s orbitals are not required by this criterion, they are neglected. Although this neglect does render the method hybridization variant, it does not affect the results to any great extent since the basis set is well defined for each atom.

As far as the one-center electron interactions are concerned, PNDO assumes all J integrals $(ii\,|\,jj)$ in the atom to be equal (irrespective of the azimuthal quantum number of i and j) and, although it specifically evaluates all k integrals $(ij|ij)$, it uses a constant value for these. In the PNDO method, the F matrix is given, as in the NDDO method for closed shell systems, by

$$F_{ii} = H_{ij}^c + \sum_B^B \sum_{kl} P_{kl} (ij|kl) - \tfrac{1}{2} \sum_{kl} P_{kl} (ik|jl) \qquad (i, j \text{ both on atom A})$$

(33)

$$F_{ij} = H_{ij} - \tfrac{1}{2} \sum_l^A \sum_k^B P_{kl} (ik|jl) \qquad (i \text{ on atom A}, j \text{ on atom B})$$

The method allows the calculation of the energy and distribution of α and β electrons separately and is thus suitable for the calculation of *open-shell systems*.

D. Semi-Empirical Evaluation of Atomic and Molecular Integrals

The three methods outlined in the last section, CNDO, (M)INDO, and PNDO, are being used as a basis for an ever-increasing number of variants, usually differing from one another by some minor change in the choice of approximation for evaluating the various integrals.

Although we give (Tables in Section E) a synopsis of the various specific approximations which have been used in the apparently most successful procedures, it might be useful at this point to discuss the most usual approximations which have or can be used in the methods. It is, however, our hope that the reader will not be tempted to combine these in yet another way and produce a further variant to the already long list of methods, but that rather it will assist him in determining which route has been used to achieve the objective of these methods, namely to reproduce the main physical properties of molecules.

1. The Correlation Problem

It is well known that if the coulomb integrals were to be evaluated directly from the expression

$$(ii|jj) = \int\int \psi_i (\mu) \, \psi_j (\nu) \, \frac{1}{r_{\mu\nu}} \, \psi_i (\mu) \, \psi_j (\nu) \, dt_\mu \, dt_\nu$$

(34)

463

using Slater orbitals for ψ_i and ψ_j, the values obtained would be too high to account satisfactorily for the experimentally observed properties. This discrepancy arises from the fact that the Hartree-Fock method is based on the premise that the motion and position of an electron occupying a given spin-orbital is independent of the motions and positions of all other electrons in the system. In reality, however, the motions and positions of electrons are interdependent (correlated). Consequently, the coulomb integrals, as calculated by Eq. (34), are too large.

This problem of electron correlation applies not only to the Hartree-Fock method, but is present in any orbital approximation.

The *correlation energy*, E_{corr}, is defined as the difference between E_{exact}, the experimentally determined ground state energy of a system, and E_{HF}, the expectation value of the Hartree-Fock operator.

$$E_{corr} = E_{exact} - E_{HF} \qquad (35)$$

Thus E_{corr} not only includes the interdependence of electron position but all other contributions that have been omitted from the Hartree-Fock Hamiltonian, e.g. relativistic effects.

Clementi[36] has shown that for a pair of electrons in a $(2p)^2$ configuration, the correlation energy amounts to between 1 and 2 e.v.

In most semi-empirical methods, the correlation energy is partially offset by replacing the actual coulomb integrals by some empirical expressions. These are designed in such a way as to reproduce experimental data in limiting cases and can hopefully be interpolated. The general framework of the methods, however, remains essentially similar to the ab initio Hartree-Fock procedures.

Thus one-center integrals can be estimated by comparing calculated and experimental energies for a sufficiently large number of appropriate states of the isolated atom.

In evaluating the two-center two-electron integrals it is assumed that the coulombic repulsion is a smooth function of internuclear distance as shown in the diagram below (Fig. 1). Hence at $R_{ij} = 0$ we have the value of the corresponding one-center two-electron integral, and at large values of R_{ij} the integrals will correspond to e^2/R_{ij}, the classical expression for the coulombic repulsion between two point-charges.

Although in the course of parameterizing the Pariser-Parr-Pople type of calculations, various methods have been developed to reproduce such a curve, not all of these have been employed in the parameterization of all valence electrons calculations. We shall now discuss the most common of these methods.

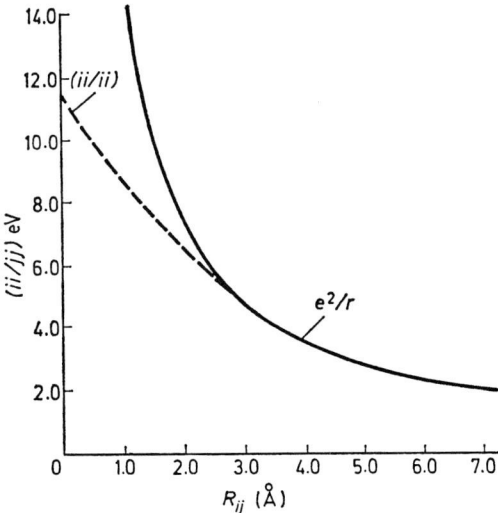

Fig. 1. Plot of the two-center two-electron coulomb integral as a function of inter-nuclear distance. - - - - Interpolated value of $<ii|jj>$

2. One-Center Integrals

At the CNDO level the one-center two-electron integrals

$$(ii|jj) = (ii|ii) = \Gamma_{AA} \qquad (36)$$

are approximated by the analytical value of the electrostatic repulsion energy of two electrons in a Slater s orbital, and this irrespective of the fact that i or j may also be p orbitals.

Pople et al. [26] pointed out in their initial (CNDO/1) scheme that since the overlap between any pair of orbitals ψ_i and ψ_j on the same atom A is set to zero, all electronic states resulting from a given configuration $(2s)^m (2p)^n$ say, of an atom or ion of a first-row element will have the same energy.

$$E(A, 2s^m 2p^n) = mU_{2s2s} + nU_{2p2p} + \tfrac{1}{2}(m+n)(m+n-1)\Gamma_{AA} \qquad (37)$$

where U_{ii} is the one-center core electron attraction integral.

This energy is thus evaluated as the weighted mean of the experimental energies of all states arising from the particular configuration.

465

Thus, in order to evaluate the parameter U_{2s2s}, say, one utilizes the ionization process represented in the equation:

$$I_{ss}(A, 2s^m 2p^n) = E(A^+, 2s^{m-1} 2p^n) - E(A, 2s^m 2p^n)$$
$$= -U_{2s2s} - (m+n-1) \Gamma_{AA} \tag{38}$$

Hence in general terms

$$U_{tt} = -I_A - (Z_A - 1) \Gamma_{AA} \tag{39}$$

Subsequently (CNDO/2) Pople suggested that since U_{tt} is essentially an atomic term it would be better approximated by the average of the ionization potential (I) and the electron affinity (A) given by

$$-A_{tt} = U_{tt} + Z_A \Gamma_{AA} \tag{40}$$

thus

$$U_{tt} = -\tfrac{1}{2}(I_A + A_A) - (Z_A - \tfrac{1}{2}) \Gamma_{AA} \tag{41}$$

Pople argued that this would better account for the tendency of an atomic orbital to both lose and gain electrons and hence *"would better represent the departure of an atom from neutrality in a molecular environment"*. The value of the orbital electronegativity $\tfrac{1}{2}(I+A)$ is determined, as described above, from appropriate spectroscopic data.

Yonezawa et al. [32a)] in a CNDO level method have approximated the one-center coulomb integrals as

$$(ii|ii) = I_t - A_t \tag{42}$$

and U_{tt} as

$$U_{tt} = -I_t - (N_t - 1)(ii|ii) - \sum_{j \neq t} N_j (ii|jj) \tag{43}$$

where

$$(ii|jj) = \tfrac{1}{2}\{(ii|ii) + (jj|jj)\} \tag{44}$$

Here N_t and N_j denote the number of electrons occupying the valence atomic orbitals ψ_t and ψ_j centered on the same atom and $\sum_{j \neq t}$ denotes summation over all-valence atomic orbitals except ψ_t, so that

$$N_t + \sum_{j \neq t} N_j = Z_A \tag{45}$$

At the INDO level Pople *et al* [27] have expressed the one-center repulsion integrals in terms of Slater-Condon F^k and G^k parameters. Thus

$$(ss|ss) = (ss|xx) = F^0 = \Gamma_{AA}$$

$$(sx|sx) = 1/3\ G^1$$

$$(xy|xy) = 3/25\ F^2 \tag{46}$$

$$(xx|xx) = F^0 + 4/25\ F^2$$

$$(xx|yy) = F^0 - 2/25\ F^2$$

Similar expressions are used for $(ss|zz)$, etc. Except for the integral F^0, evaluated from Slater orbitals, semi-empirical values are used for G^1 and F^2 chosen to give the best fit with atomic spectra. At this level Pople thus expresses the energy of the average state associated with the configuration $(2s)^m (2p)^n$ as

$$E(m, n) = mU_{ss} + nU_{pp} + \tfrac{1}{2}(m+n)(m+n-1)F^0$$
$$- \tfrac{1}{6}mnG^1 - \tfrac{1}{25}n(n-1)F^2 \tag{47}$$

He then deduces the following relationships between $\tfrac{1}{2}(I+A)$, the orbital electronegativity, and U_{tt}:

Hydrogen

$$-\tfrac{1}{2}(I+A)_s = U_{ss} + \tfrac{1}{2}F^0$$

Lithium

$$-\tfrac{1}{2}(I+A)_s = U_{ss} + \tfrac{1}{2}F^0$$
$$-\tfrac{1}{2}(I+A)_p = U_{pp} + \tfrac{1}{2}F^0 - \tfrac{1}{12}G^1$$

Beryllium

$$-\tfrac{1}{2}(I+A)_s = U_{ss} + \tfrac{3}{2}F^0 - \tfrac{1}{2}G^1$$
$$-\tfrac{1}{2}(I+A)_p = U_{pp} + \tfrac{3}{2}F^0 - \tfrac{1}{4}G^1$$

Boron to Fluorine

$$-\tfrac{1}{2}(I+A)_s = U_{ss} + (Z_A - \tfrac{1}{2})F^0 - \tfrac{1}{6}(Z_A - \tfrac{3}{2})G^1$$
$$-\tfrac{1}{2}(I+A)_p = U_{pp} + (Z_A - \tfrac{1}{2})F^0 - \tfrac{1}{3}G^1 - \tfrac{2}{25}(Z_A - \tfrac{5}{2})F^2$$

Yonezawa et al. [32 b)], in an INDO level method, by introducing one-center exchange integrals have approximated U_{tt} as

$$U_{tt} = -I_t - (N_t - 1)(ii|ii) - \sum_{j \neq t} N_j \{(ii|jj) - \tfrac{1}{2}(ij|ij)\} \qquad (48)$$

They approximate the one-center exchange integrals as

$$(sp|sp)_A = 0.045 Z_A (ss|pp)_A$$

$$(pp^1|pp^1)_A = 0.011 Z_A (pp|p^1p^1)_A \qquad (49)$$

where p and p^1 are two different p orbitals situated on atom A.

Dewar et al. [35)] in their MINDO method follow Pople's approach and express the one-center electron repulsion integrals as in Eq. (46). Although they use the values quoted by Pople for G^1 and F^2, however, they evaluate U_{tt} and F^0 for each atom having the ground state configuration $s^n p^m$, from transition energies among the high-spin states of the configurations $s^n p^{m+1}$, $s^n p^m$, $s^n p^{m-1}$ and $s^{n-1} p^{m+1}$.

At the PNDO level Dewar and Klopman have assumed that the repulsion between two valence shell electrons of an atom is the same and is independent of the orbitals occupied. This repulsion has one of two values, A^+ and A^-, depending upon whether the electron spins are paired or opposed. The repulsion integrals can then be expressed in terms of these quantities as

$$(ii|jj) = A^- \qquad (i = j \text{ or } i \neq j)$$

$$(ij|ij) = A^- - A^+ \qquad (i \neq j) \qquad (50)$$

The atomic terms U and A are determined from the atomic spectra of the corresponding isolated atom by selecting only those states which are of importance for bonding in molecules. Klopman [24)] has achieved this by using barycenters of states correlated by means of a simplified Slater-Condon type equation of the form

$$E = \sum_i U_i^l + \tfrac{1}{2}\sum_i\sum_{j \neq i} A_{ij}^+ \delta_{ij} + \tfrac{1}{2}\sum_i\sum_{j \neq i} A_{ij}^- (1 - \delta_{ij}) \qquad (51)$$

where the superscript l indicates U's dependence on the orbital quantum numbers and δ_{ij} is a Kroneker delta equal to 1 when the spins of the electrons occupying orbitals ψ_i and ψ_j are paired, and 0 otherwise.

3. Two-Center Two-Electron Integrals

The most difficult problem encountered in the design of semi-empirical quantum mechanical methods is the determination of a satisfactory way of calculating the two-center two-electron integrals.

The general guidelines are, as stated above, that the expression reduces to that for one-center two-electron integrals at zero distance, and tends toward e^2/R at large distances (see Fig. 1).

Many such expressions have been suggested, most of them in the context of the Pariser-Parr-Pople calculations for π conjugated species. We describe below most of the approximations which were found suitable even though some of them have not been used in the methods designed for σ bonded molecules.

a) The Uniformly Charged Sphere Method

In this approach, due to Parr [37], each p atomic orbital containing a pair of electrons is replaced by two tangentially touching, non-conducting charged spheres of diameter R_A given by

$$R_A = \left(\frac{4.597}{Z_A}\right) \times 10^{-8} \text{ cm} \tag{52}$$

where Z_A is the Slater effective nuclear charge of atom A. In effect this model places a point charge $e/2$ at a distance $4.34\ a_0/Z_A$ above and below the position of a $2p_z$ Slater atomic orbital and $9.1\ a_0/Z_A$ above and below the position of a $3p_z$ Slater atomic orbital, where e is the electronic charge. The $(ii|jj)$ integral is then equated to the repulsion, evaluated by classical electrostatics, between these spheres. This is shown to be:

$$(ii|jj) = \frac{e^2}{2\,r}\left[\frac{1}{\sqrt{1+(1/2\,r)^2\,(R_A-R_B)^2}} + \frac{1}{\sqrt{1+(1/2\,r)^2\,(R_A+R_B)^2}}\right] ev \tag{53}$$

where r is the internuclear distance. At distances of less than 2.80 Å the charged sphere model breaks down, since overlap would now occur. In this case Pariser and Parr joined the values of $(ii|jj)$ extrapolated to $r_{ij}=0$ with those of the charged sphere model at $r \geq 2.80$ Å. For this they used the equation

$$ar + br^2 = \tfrac{1}{2}\left[(ii|ii) + (jj|jj)\right] - (ii|jj) \tag{54}$$

where the parameters a and b are determined from the charged sphere model at the fixed internuclear distances of 2.80 Å and 3.70 Å respectively. Thus the two-center integrals are given by

$$(ii|jj) = \tfrac{1}{2}\left[(ii|ii) + (jj|jj)\right] - ar - br^2 \tag{55}$$

This method has been extensively used in Pariser-Parr-Pople calculations on conjugated systems. It has also been employed by Jaffe [31] in his parameterization of CNDO/2.

b) The Split p Orbital Method

Dewar *et al.*, in a series of papers [39] developed the split p orbital method as a means of taking into account "vertical correlation," i.e. the tendency of the position of one of a pair of electrons in a p or π type orbital to predetermine the position of the second electron by keeping their mutual distance as large as possible. The p or π-orbital is split into two parts along its nodal plane and each half is treated as a separate orbital. Thus Dewar has expressed the two-center two-electron integral as

$$(ii|jj) = \frac{e^2}{\sqrt{r_{AB}^2 + 4\,R^2}} \tag{56}$$

i.e. the interaction between two electrons occupying lobes on opposite sides of the nodal plane (R being the radius of one of the tangentially touching spheres used to simulate a p orbital).

Dewar's method has received considerable criticism, and has not been used in all-valence electrons calculations.

c) Mataga-Nishimoto Method

In this method [39] the two-center repulsion integrals are calculated from the corresponding one-center repulsion integrals by means of the empirical relationship

$$(ii|jj) = e^2/R \tag{57}$$

where $R = (a + r_{AB})$ and r_{AB} is the internuclear distance between atoms A and B.

The parameter a is evaluated from valence state ionization potentials I_i and electron affinity A_i in the same valence state. Two distinct cases may arise:

a) Homonuclear case

$$a = \frac{e^2}{(I_i - A_i)} \tag{58}$$

b) Heteronuclear case

$$a = \frac{e^2}{1/2\,[(I_i - A_i) + (I_j - A_j)]} \tag{59}$$

d) Ohno Method

Ohno [40] has suggested the following empirical relationship which he applied to Pariser-Parr-Pople calculations

$$(ii|jj) = e^2/R \tag{60}$$

where $R = (a^2 + r_{AB}^2)^{1/2}$ and a is given as in the Mataga Nishimoto method, by

$$a = e^2/(I - A) \tag{61}$$

Yonezawa et al. [32] in their CNDO and INDO level methods have adopted such an approximation for their evaluation of the two-center two-electron integrals.

e) Klopman Method

Klopman [24], in an all-valence electron treatment of small molecules, suggested the following relationship

$$(ii|jj) = \frac{e^2}{\sqrt{r_{AB}^2 + (\varrho_A + \varrho_B)^2}}\, e.v. \tag{62}$$

where $\varrho_A = e/2\,A_i^-$ and $\varrho_B = e/2\,A_j^-$. In both their MINDO/1 and /2 methods Dewar et al. [34,35] have used the Ohno-Klopman expression.

Table 3. *Calculation of coulomb interactions between electrons in carbon π orbitals (e. V.)*

Method	Distance, Å			
	0	1.397	2.420	2.794
Slater	16.93	9.027	5.668	4.968
Pariser-Parr	10.53—11.1	7.30	5.46	4.90
Dewar	10.02	7.61	—	—
Klopman	11.144	7.56	5.25	4.68

f) Dewar and Klopman Method

Dewar and Klopman [29] in their PNDO method have assumed different values for the integral $<ii|jj>$ depending on the nature of the orbitals involved and their mode of overlap. The two-center two-electron integrals arising in their method fall into three groups according to the correlation energy involved

a) large correlation $(s\sigma|s\sigma)$ $(p\pi|p\pi)$ $(s\sigma|p\pi)$

b) medium correlation $(s\sigma|p\sigma)$ $(p\sigma|p\pi)$

c) small correlation $(p\sigma|p\sigma)$

This same grouping also applied to the integrals when they were estimated theoretically using Slater-Zener orbitals. They therefore developed expressions for the integrals that would duplicate this grouping, subject to the condition that the integral $(ii|jj)$ between orbitals of two identical atoms should converge to the corresponding one-center integral $(ii|ii)$ at zero internuclear separation.

Group I

$$(ii|jj) = e^2\{r_{ij}^2 + (\varrho_i + \varrho_j)^2\}^{-1/2}$$

Group II

$$(ii|jj) = e^2\{r_{ij}^2 + (\varrho_i T_{ij} + \varrho_j)^2\}^{-1/2}$$

$$(63)$$

where ψ_i is a $p\sigma$ atomic orbital

Group III

$$(ii|jj) = e^2\{r_{ij}^2 + (\varrho_i + \varrho_j)^2 T_{ij}^2\}^{-1/2}$$

where both ψ_i and ψ_i are $p\sigma$ orbitals.

All orbitals are s or $p \pi$ unless otherwise stated and

$$T_{ij} = \exp\left[-r_{ij}/2\,(\varrho_i + \varrho_j)\right]$$

$$(64)$$

The values calculated in this manner for carbon atoms at internuclear distance of 1.55 Å are compared in Table 4 to those calculated from Slater orbitals.

g) Pople Method

Pople et al. [26,27] in their CNDO and INDO methods evaluate the two-center two-electron integrals

$$(ii|jj) = \Gamma_{AB}$$

$$(65)$$

Table 4. *Values for carbon-carbon two-center repulsion integrals*

Group	Type	Value of integral, ev	
		Calcd. using Slater-Zener AO's	Calcd. from Dewar-Klopman expressions
1	$s\sigma:s\sigma$	9.28	
	$s\sigma:p\pi$	9.12	7.13
	$p\pi:p\pi$	8.98	
	$p\pi(x):p\pi(y)$	8.98	
2	$s\sigma:p\sigma$	9.61	7.81
	$p\pi:p\sigma$	9.41	
3	$p\sigma:p\sigma$	9.99	8.45

which represent the interaction between electrons in any valence atomic orbitals on atoms A and B, as the two-center coulomb integral

$$\Gamma_{AB} = (s_A s_A | s_B s_B)$$
$$= \iint s_A^2(\mu)\ 1/r_{\mu s}\ s_B^2(\nu)\ dt_\mu\ dt_\nu \tag{66}$$

involving exclusively Slater s orbitals.

4. Coulomb Penetration Integrals

In the CNDO/1 method the penetration terms $(Z_B \Gamma_{AB} - v_{AB})$ were evaluated by approximating all coulomb penetration integrals v_{AB} as

$$v_{AB} = \int s_A^2(\mu)\ \frac{Z_B}{r_{\mu B}}\ d\tau_\mu \tag{67}$$

where Z_B is the core charge on atom B and $r_{\mu B}$ the distance of the electron μ from B. s_A is the Slater $2s$ orbital of atom A. A major failure of CNDO/1 was, however, its inability to give reasonable values for *bond lengths* (these were too short) and *bond energies* (these were too large) for diatomic molecules.

Pople *et al.* attempted to correct this in their CNDO/2 method by neglecting altogether the *penetration terms*, i.e. by setting $v_{AB} = Z_B \Gamma_{AB}$. They argued that this was a legitimate modification since neglect of overlap distribution introduces errors, similar to but opposite in sign to neglect of penetration. Although bond distances were improved, the bond dissociation energies remained too large.

Yonezawa *et al.* in their CNDO level method approximate the penetration integral $v_{AB} = (B|ii)$ be setting them equal to the negative of the corresponding repulsion integral, thus

$$(B|ii) = -Z_B \, (s_B s_B|ii) \tag{68}$$

However, in their INDO level method they use the expression

$$(B|ii) = -\sum_k^B N_k \, (kk|ii) \tag{69}$$

The policy of neglecting penetration terms has been continued by Pople *et al.* in their INDO method and by Dewar *et al.* in their MINDO method.

Dewar and Klopman (PNDO) approximate the *penetration integral* v_{iB} as

$$v_{iB} = \Gamma_{i(p\sigma)_B} + (Z_B - 1) \, \Gamma_{i(p\pi)_B} \tag{70}$$

where Z_B is the charge on the core of atom B. The $\Gamma_{i(p\sigma)_B}$ term denotes the interaction between an electron in a given orbital ψ_i on atom A and a positive hole located in the $p\sigma$ orbital of the core of atom B, and $\Gamma_{i(p\pi)_B}$ the interaction between an electron in ψ_i and a positive hole located in the s or p-π orbital on the core of atom B. For cores other than hydrogen $(Z_B > 1)$ the core-electron interaction is set equal to the average interaction of the electron in orbital ψ_i and a positive hole located in every orbital of atom B. For hydrogen $(Z_B = 1)$ the core-electron interaction is set equal to, but opposite in sign to, the interaction with an electron situated in the hydrogen atom orbital.

5. The Two-Center One-Electron "Resonance" Integral

The one-electron resonance integral H_{ij} can be interpreted physically as the *energy of an electron occupying the overlap cloud* between the atomic orbitals ψ_i and ψ_j and moving in the field of the core and remaining electrons.

In the CNDO and INDO methods Pople *et al.* set the resonance integral directly proportional to the overlap integral S_{ij} between the orbitals ψ_i and ψ_j centered on atoms A and B

$$H_{ij} = \beta^0_{AB} S_{ij} \tag{71}$$

where Slater atomic orbitals are used to calculate S_{ij}. In order that the calculations should be rotational invariant the parameter β_{AB}^0 should be characteristic of ψ_i and ψ_j but independent of their positions in space. To this end Pople suggested using the average of the β parameter for each atom.

$$\beta_{AB}^0 = \tfrac{1}{2} (\beta_A^0 + \beta_B^0) \qquad (72)$$

Here β_A^0 and β_B^0 are adjustable, empirically determined parameters, chosen so that the calculation gives the best fit between CNDO and ab initio L.C.A.O. S.C.F. calculated charges in selected diatomic molecules.

Dewar and Klopman pointed out that H_{ij} may be expected to be proportional to the magnitude of the overlap cloud, i.e. to the overlap integral S_{ij}, and to some mean of the attraction energy experienced by an electron in the overlap region.

$$H_{ij} = \beta_{AB} \, S_{ij} \, (I_i + I_j) \, \{R_{AB}^2 + (\varrho_i + \varrho_j)^2\}^{-1/2} \qquad (73)$$

Here I_i and I_j are the valence-state ionization potentials of the atomic orbitals ψ_i and ψ_j, calculated for the appropriate barycenters. R_{AB} is the internuclear distance between the atoms of which ψ_i and ψ_j are atomic orbitals. S_{ij} is calculated using Slater orbitals. The quantity $[R_{AB}^2 + (\varrho_i + \varrho_j)]^{-1/2}$ is essentially equal to Γ_{ij} (Eq. 62).

In order to reduce the number of parameters in the PNDO treatment, they assumed the *empirically determined parameter* β_{AB} had a common value for orbitals of two atoms A and B, regardless of the type of orbitals (s or p) or the mode of overlap (σ or π) and that

$$\beta_{AB} = \sqrt{\beta_{AA} \, \beta_{BB}} \qquad (74)$$

Although the Dewar-Klopman expression is more complicated than the resonance expressions so far discussed, they point out that attempts to use simpler expressions resulted in less success. Thus, the omission of the terms I_i and I_j gave results for unsaturated molecules such as ethylene in which the orbital energies appeared in the wrong orders, while the omission of the term in R resulted in incorrect heats of formation.

Dewar *et al.* in their MINDO/1 method found it more convenient to approximate H_{ij} for two atoms A and B at an internuclear distance R_{AB} by

$$H_{ij} = S_{ij} \, [I_i + I_j] \left(A_{AB} + \frac{B_{AB}}{R_{AB}^2} \right) \qquad (75)$$

where I_i and I_j, the neutral-atom valence-state ionization potentials, are approximated by

$$I_s^A = U_{ss}^A + (Z_A - 1) (F_A^0 - \tfrac{1}{6} G_A^1)$$
$$I_p^A = U_{pp}^A + \tfrac{1}{3} (Z_A - 1) (3 F_A^0 - \tfrac{1}{6} G_A^1 - 0.28 F_A^2) \tag{76}$$

The parameters A_{AB} and B_{AB} are estimated by fitting the observed heats of formation of suitable reference compounds. In their most recent (MINDO/2) method, Dewar *et al.* have expressed the resonance integral in the form

$$H_{ij} = \beta_0 S_{ij} (I_i + I_j) \tag{77}$$

Here the parameter β_0 is determined by a least squares fit to the heats of formation of selected bonds in a group of standard molecules chosen to include all types of hydrocarbons. As in the Pople methods, S_{ij} is determined from Slater orbitals.

Yonezawa *et al.*, in their INDO level method, adopted the following expression for the resonance integral

$$H_{ij} = \frac{S_{ij}}{2} [-(Z_A + Z_B)/R_{AB} - (B/ii) - (A/jj) + H_{ii} + H_{jj}] \tag{78}$$

6. Core-Core Repulsions

Pople *et al.* in their CNDO and INDO methods and Yonezawa *et al.* in their INDO level approximation set C_{AB}, the core-core repulsion, equal to a point charge potential, thus:

$$C_{AB} = Z_A Z_B R_{AB}^{-1} \tag{79}$$

where Z_A and Z_B are the respective core charges.

Such an approximation, however, results in calculated binding energies which are too small. The reason for this may lie in the fact that the potential field in which the electrons move in a molecule is greater than that in the isolated atom; consequently, the atomic orbitals in the molecule will be more compact. In the semi-empirical treatments discussed above, however, "atomic" parameters are determined from spectroscopic data for isolated atoms. Such approaches therefore assume that the effective nuclear charge is the same whether the atom be isolated or bound in a molecule. It may thus be that in order to get realistic binding energies, either we abandon this assumption or make some allowance for it. The latter can conveniently be achieved by modifying the form of the core-core repulsion equation. Its advantage is that it does not affect the calculations of the electron distribution or orbital energies. Klopman and Dewar in their PNDO method thus calculated the core-core repulsion by meens of a parametric function. The function is chosen to satisfy two boundary conditions. For large R_{ij} it must approach the corresponding interelectronic repulsion between neutral atoms in order that the net potential due to a neutral atom should vanish at large distances, while for small R_{ij} it must have a value between this, and that calculated for point charges. After trying a large number of one-parameter functions they found the most successful one to be

$$C_{AB} = E_{AB} + [Z_A Z_B e^2 / R_{AB} - E_{AB}] e^{-\alpha_{AB} R_{AB}} \qquad (80)$$

C_{AB} is the core repulsion between atoms A and B;

E_{AB} is the corresponding electronic repulsion between neutral atoms A

and B (i.e., $(ii|jj)$ summed over all occupied valence orbitals);

Z_A and Z_B are the formal core charges in units of e (i.e. the number of valence electrons) of the two atoms;

α_{AB} is a parameter.

In order to reduce the number of parameters in the treatment, they assumed that the value of α_{AB} for two dissimilar atoms A and B is given in terms of the parameters α_{AA} and α_{BB} for pairs of similar atoms by

$$\alpha_{AB} = \sqrt{\alpha_{AA} \alpha_{BB}} \qquad (81)$$

Dewar *et al.* adopted a similar function in the MINDO/2 method.

E. Tables

We give here a synopsis of the various methods of approximation together with their objectives, areas where they have found their most successful application, e.g. dipole moments, their field of application, i.e. types of compounds to which they have been applied, and their limitations. In cases where two methods have been published, e.g. CNDO/1 and CNDO/2, we give the second improved method. Where a program for the method is available from "Quantum Chemistry Program Exchange (QCPE), Chemistry Department, Room 204, Indiana University, Bloomington, Indiana 47401," we list as a source the QCPE program number.

Method I	CNDO/2

Authors   J. A. Pople and G. A. Segal [26]

a) Scope of method

Objectives   Charge densities

Compounds   Hydrocarbons, molecules of the form AB_2 and AB_3, organic compounds containing hetroatoms (N, O, F), nitrogen and oxygen hetrocycles, flurinated benzenes and nitrobenzenes

Successes   Bond angles, dipole moments, bending force constants, bond length, n.m.r. correlation

Failures   Heats of formation, ionization potentials, and electron affinities (both too large by several electron volts,) spectral transitions (too high an energy), total energy and energies of occupied orbitals are uniformly too negative. Virtual orbitals lie too high in energy

Source   QCPE No. 91, 141, 142, 144

b) Approximations used

One-center one-electron $U_{ii} = -1/2 \{I_i + A_i\} - \{Z_A - 1/2\} \Gamma_{AA}$

One-center two-electron $\Gamma_{AA} = \iint s_A^2(\mu) \, 1/r_{\mu\nu} \, s_A^2(\nu) \, d\tau_\mu \, d\tau_\nu$

Penetration $V_{AB} = Z_B \Gamma_{AB}$

Resonance $\qquad H_{ij} = \beta^0_{AB} \, S_{ij}$

$\qquad\qquad\qquad\qquad\qquad \beta^0_{AB} = \frac{1}{2} \, (\beta^0_A + \beta^0_B)$

Two-center two-electron $\quad \Gamma_{AB} = \iint s^2_A \, (\mu) \, 1/r_{\mu\nu} \, s^2_B \, (\nu) \, d\tau_\mu \, d\tau_\nu$

Core-core repulsion $\qquad C_{AB} = Z_A Z_B R^{-1}_{AB}$

c) Matrix elements

$$F_{ii} = -1/2 \, (1_i + A_i) + \{(P_{AA} - Z_A) - 1/2 \, (P_{ii} - 1)\} \, \Gamma_{AA} + \sum_{B \neq A} (P_{BB} - Z_B) \, \Gamma_{AB}$$

$$F_{ij} = 1/2 \, \{\beta^0_A + \beta^0_B\} \, S_{ij} - 1/2 \, P_{ij} \, \Gamma_{AB}$$

d) Total electronic energy

$$E = \sum_A E_A + \sum_{A<B} \sum E_{AB}$$

where

$$E_A = \sum_i^A P_{ii} \, U_{ii} = 1/2 \sum_i^A \sum_j^A (P_{ii} \, P_{jj} - 1/2 \, P^2_{ij}) \, \Gamma_{AA}$$

and

$$E_{AB} = \sum_i^A \sum_j^B \{2 \, P_{ij} \, \beta^0_{AB} \, S_{ij} - 1/2 \, P^2_{ij} \, \Gamma_{AB}\} + \{Z_A Z_B R^{-1}_{AB} - P_{AA} \, V_{AB} - P_{BB} \, V_{AB}$$
$$+ P_{AA} \, P_{BB} \, \Gamma_{AB}\}$$

Method II	CNDO/2 Level

Authors   G. Del Bene and H. H. Jaffe [31]

a) Scope of method

Objectives   Spectra

Compounds   Aromatic derivatives and heterocycles

Successes   Spectra of conjugated systems

Failures   Heats of formation, molecular geometries

Source

b) Approximations used

One-center one-electron $U_{ii} = -\frac{1}{2}(I_i + A_i) - (Z_A - \frac{1}{2})\, \Gamma_{AA}$

One-center two-electron $\Gamma_{AA} = I_i - A_i$

Penetration $V_{AB} = Z_B\, \Gamma_{AB}$

Resonance $H_{ij} = \frac{1}{2} K\, (\beta_A^0 + \beta_B^0)\, S_{ij}$

$K = 1.0 \quad$ for σ bonds

$= 0.585$ for π bonds

Two-center two-electron $\Gamma_{AB} =$ charged sphere approximation

Core-core repulsion $C_{AB} = Z_A\, Z_B\, R_{AB}^{-1}$

c) *Matrix elements*

$$F_{ii} = -\tfrac{1}{2}\,(I_i + A_i) + [(P_{AA} - Z_A) - \tfrac{1}{2}\,(P_{ii} - 1)]\,\Gamma_{AA} + \sum_{B \neq A} (P_{BB} - Z_B)\,\Gamma_{AB}$$

$$F_{ij} = \tfrac{1}{2}\,K\,[\beta_A^0 + \beta_B^0)\,S_{ij} - \tfrac{1}{2}\,P_{ij}\,\Gamma_{AB}$$

d) *Total electronic energy*

$$E = \sum_A E_A + \sum_{A<B} \sum E_{AB}$$

where

$$E_A = \sum_i^A P_{ii}\,U_{ii} = 1/2 \sum_i^A \sum_j^A (P_{ii}\,P_{jj} - 1/2\,P_{ij}^2)\,\Gamma_{AA}$$

and

$$E_{AB} = \sum_i^A \sum_j^B \{2\,P_{ij}\,\beta_{AB}^0\,S_{ij} - 1/2\,P_{ij}^2\,\Gamma_{AB}\} + \{Z_A\,Z_B\,R_{AB}^{-1} - P_{AA}\,V_{AB} - P_{BB}$$

$$V_{BA} + P_{AA}\,P_{BB}\,\Gamma_{AB}\}$$

Additional feature: A limited configuration interaction calculation is performed

Method III	INDO

a) Scope of method

Authors	J. A. Pople, D. L. Beveridge and P. A. Dobosh [27]

a) Scope of method

Objectives	Spin densities
Compounds	Hydrocarbons, molecules of the form AB_2 and AB_3, organic compounds containing hetroatoms (N, O, F), free radicals
Successes	Spin densities, hyperfine coupling constants, molecular geometries
Failures	Heats of formation, ionization potentials, electron affinities, spectral transitions
Source	QCPE No. 141, 142, 144

b) Approximation used

One-center one-electron

Hydrogen

$$-1/2\,(I+A)_s = U_s + 1/2\,\Gamma_{\text{HH}}$$

Lithium

$$-1/2\,(I+A)_s = U_s + 1/2\,F^0$$
$$-1/2\,(I+A)_p = U_p + 1/2\,F^0 - 1/12\,G^1$$

Berylium

$$-1/2\,(I+A)_s = U_s + 3/2\,F^0 - 1/2\,G^1$$
$$-1/2\,(I+A)_p = U_p + 3/2\,F^0 - 1/4\,G^1$$

Boron to Fluorine

$$-1/2\,(I+A)s = U_s + (Z_A - 1/2)\,F^0$$
$$-1/6\,(Z_A - 3/2)\,G^1$$

$$-1/2\,(I+A)_p = U_p + (Z_A - 1/2)\,F^0$$
$$-1/3\,G^1 - 2/25\,(Z_A - 5/2)\,F^2$$

One-center two-electron

$$(ss \mid ss) = (ss \mid xx) = F^0 = \Gamma_{AA}$$

$$(sx \mid sx) = 1/3 \ G^1$$

$$(xy \mid xy) = 3/25 \ F^2$$

$$(xx \mid xx) = F^0 + 4/25 \ F^2$$

$$(xx \mid yy) = F^0 - 2/25 \ F^2$$

Penetration

$$V_{AB} = Z_B \ \Gamma_{AB}$$

Resonance

$$H_{ij} = \beta^0_{AB} \ S_{ij} \qquad \beta^0_{AB} = \tfrac{1}{2} \ (\beta^0_A + \beta^0_B)$$

Two-center two-electron

$$\Gamma_{AB} = (s_A \ s_A \mid s_B s_B)$$

$$= \iint s^2_A \ (\mu) \ 1/r_{\mu\nu} \ s^2_B \ (\nu) \ dt_\mu \ dt_\nu$$

Core-core repulsion

$$C_{AB} = Z_A \ Z_B \ R^{-1}_{AB}$$

c) Hartree-Fock matrix elements

$$F^a_{ii} = U_{ii} + \sum_j^A \{P_{ii} \ (ii \mid jj) - P^a_{jj} \ (ij \mid ij)\} + \sum_{B \neq A} (P_{BB} - Z_B) \ \Gamma_{AB}$$

$$(i \text{ on atom } A)$$

$$F^a_{ik} = (2 \ P_{ik} - P^a_{ik}) \ (ik \mid ik) - P^a_{ik} \ (ii \mid kk)$$

$$(i \neq k \text{ both on atom } A)$$

d) Total electronic energy

$$E = \sum_A E_A + \sum_{A < B} \sum E_{AB}$$

where

$$E_A = \sum_i^A P_{ii} \ U_{ii} = 1/2 \sum_i^A \sum_j^A (P_{ii} \ P_{jj} - 1/2 \ P^2_{ij}) \ \Gamma_{AA}$$

and

$$E_{AB} = \sum_i^A \sum_j^B \{2 \ P_{ij} \ \beta^0_{AB} \ S_{ij} - 1/2 \ P^2_{ij} \Gamma_{AB}\} + \{Z_A \ Z_B \ R^{-1}_{AB} - P_{AA} \ V_{AB} - P_{BB} \ V_{BA}$$

$$+ P_{AA} \ P_{BB} \ \Gamma_{AB}\}$$

Method IV   MINDO/2

Authors   M. J. S. Dewar and E. Haselbach [35]

a) Scope of method

Objectives   Reproduce approximately a Morse curve for a molecule in the gaseous state

Compounds   Hydrocarbons

Successes   Heats of formation, molecular geometries

Failures   Spectra

Source   MINDO/1, QCPE No. 137

b) Approximations used

One-center one-electron $I_s^A = U_{ss}^A + (Z_A - 1)(F_A^0 - 1/6\, G_A^1)$

$I_p^A = U_{pp}^A + 1/3\,(Z_A - 1)\,(3\,F_A^0 - 1/6\,G_A^1 - 0.28\,F_A^2)$

One-center two-electron $(ss|ss) = (ss|pp) = F^0$

$(sp|sp) = 1/3\,G^1$

$(pp|pp) = F^0 + 4/25\,F^2$

$(pp|p^1p^1) = F^0 - 2/25\,F^2$

$(pp^1|pp^1) = 3/25\,F^2$

Penetration	$V_{AB} = Z_B \, \Gamma_{AB}$
Resonance	$H_{ij} = \beta^0 S_{ij} \, (I_i + I_j)$
Two-center two-electron	$\Gamma_{AB} = e^2/ \, (r_{AB}^2 + (\varrho_A + \varrho_B)^2)^{-1/2}$
	$\varrho_A = e/2 \, A^-$ and $\varrho_B = e/2 \, A^+$
Core-core repulsion	$C_{AB} = E R_{AB} + (P R_{AB} - E R_{AB}) \, e^{-a R_{AB}}$

c) Hartree-Fock matrix elements

$$F_{ii}^{AA} = U_{ii} + 1/2 \, q_i \, (ii|ii) + \sum_{j \neq i}^{A} q_j \, \{(ii|jj) - 1/2 \, (ij|ij)\} + \sum_{B \neq A} (Q_B - Z_B) \, \Gamma_{AB}$$

$$F_{ij}^{AA} = P_{ij} \, \{3/2 \, (ij|ij) - 1/2 \, (ii|jj)\}$$

$$F_{ij}^{AB} = H_{ij} - 1/2 \, P_{ij} \, \Gamma_{AB}$$

Here q refers to electron densities on the atomic orbitals, Q is the total valence-shell electron density for an atom, and P represents the bond order matrix.

d) Total electronic energy

$$E = 1/2 \sum_{i} \sum_{j} P_{ij} \, (H_{ij} + F_{ij}) + \sum_{A < B} \sum Z_A \, Z_B \, \Gamma_{AB}$$

Method V	INDO level

Authors	H. Kato, H. Konishi, H. Yamabe and T. Yonezawa [32b]

a) Scope of method

Objectives	Electronic structure
Compounds	Hydrocarbons, organic compounds containing heteroatoms (O, N)
Successes	Not enough data to assess
Failures	Not enough data to assess
Source	

b) Approximations used

One-center one-electron	$U_{ii} = -I_i - (N_i - 1)(ii|ii) - N_j\{(ii|jj) - 1/2\,(ij|ij)\}$
One-center two-electron	$(sp|sp)_A = 0.045\,Z_A\,(ss|pp)_A$
	$(pp^1|pp^1)_A = 0.011\,Z_A\,(pp|p^1p^1)_A$
	$(ii|ii) = (I_i - A_i)$
Penetration	$V_{AB} = \overset{on\ B}{\underset{k}{\sum}} N_k\,(kk|ii)$
Resonance	$H_{ij} = \dfrac{S_{ij}}{2}\{-(Z_A + Z_B)(C/R_{AB}) - (B|ii) - (A|jj) + H_{ii} + H_{jj}\}$

Two-center two-electron $\quad (ii|jj) = 1/\,(a^2 + R_{ij}^2)^{1/2}$

where

$$1/a = (ii|ii) = I_i - A_i$$

Core-core repulsion $\qquad C_{AB} = Z_A\, Z_B\, R_{AB}^{-1}$

c) Hartree-Fock matrix elements

$$F_{rs} = H_{rs} + \sum_{t,\,u} P_{tu}\, [<rs|tu> - \tfrac{1}{2} <rt|su>] \qquad \text{[Roothaan's equation]}$$

d) Total electronic energy

$$E = E_e + \sum_{A>B} \sum Z_A\, Z_B / R_{AB}$$

where

$$E_e = 1/2 \sum_{ij} P_{ij}\, (H_{ij} + F_{ij})$$

Method VI   PNDO

Authors
M. J. S. Dewar and G. Klopman [28)]

a) Scope of method

Ojectives
Heats of formation

Compounds
Hydrocarbons

Successes
Heats of formation and ionization potentials

Failures
Bond distances and spectra

Source

b) Approximations used

One-center one-electron
$$U_{ii} = W_i^{(M)} + P_i^\beta A_M^- + \sum_{j \neq i}^M (P_{jj}^a A_M^+ + P_{jj}^\beta A_M^-)$$

One-center two-electron
$$(ii|jj) = A^- \quad (i = j \text{ or } i \neq j)$$
$$(ij|ij) = A^- - A^+ \quad (i \neq j)$$

Penetration
$$V_{iB} = \Gamma_{i(p\sigma)_B} + (Z_B - 1) \Gamma_{i(p\pi)_B}$$

Resonance
$$H_{ij} = (\beta_{ij}) S_{ij} (I_i + I_j) \{R_{ij}^2 + (\varrho_i + \varrho_j)\}^{-1/2}$$
$$\beta_{ij} = (\beta_{ii} \beta_{jj})^{1/2}$$

Two-center two-electron

    Group I
$$\gamma^{\pi\pi} = (ii|jj) = e^2 \{r_{ij}^2 + (\varrho_i + \varrho_j)^2\}^{-1/2}$$
    Group II
$$\gamma^{\sigma\pi} = (ii|jj) = e^2 \{r_{ij}^2 + (\varrho_i T_{ij} + \varrho_j)^2\}^{-1/2}$$

Group III
$$\gamma^{\sigma\sigma} = (ii|jj) = e^2 \left\{ r_{ij}^2 + (\varrho_i + \varrho_{jj})^2 \ T_{ij}^2 \right\}^{-1/2}$$

where:
$$T_{ij} = e^{-r_{ij}/2(\varrho_i + \varrho_j)}$$

Core-core repulsion
$$C_{AB} = E_{AB} + [Z_A \ Z_B \ e^2/R_{AB} - E_{AB}] \ e^{-\alpha_{AB}R_{AB}}$$

$$\alpha_{AB} = \sqrt{\alpha_{AA} \ \alpha_{BB}}$$

c) Hartree-Fock matrix elements

$$F_{kk} = W_k^{(M)} + \sum_{N \neq M} [k_x^2 \ V_{kN}^\sigma + (k_y^2 + k_z^2) V_{kN}^\pi] + q_k^\beta A_M^-$$

$$+ \sum_{m \neq k}^{(M)} (q_m^\alpha A_M^+ + q_m^\beta A_M^-) + \sum_l^{(N)} \sum_n^{(N)} (p_{ln}^\alpha + p_{ln}^\beta) \ (kk, \ ln)$$

$$F_{km}^{(M)} = \sum_{N \neq M} [k_x m_x V_{kN}^\sigma + (k_y m_y + k_z m_z) V_{kN}^\pi] - p_{km}^\alpha A_M^+ + \sum_l^{(N)} \sum_n^{(N)} (p_{ln}^\alpha + p_{ln}^\beta) \ (lm, \ ln)$$

$$F_{kl}^{(M,N)} = k_x \ l_x \ \beta_{kl}^\sigma + (k_y \ l_y + k_z \ l_z) \ \beta_{kl}^\pi - \sum_m^{(M)} \sum_n^{(N)} p_{mn} \ (km, \ ln)$$

d) Total electronic energy

$$E = \sum_k^{(M)} ((q_k^\alpha + q_k^\beta) \ W_k$$

$$+ \sum_m^{(M)} \left\{ \sum_{N \neq M} [k_x \ m_x \ V_{kN}^\sigma + (k_y \ m_y + k_z \ m_z) \ V_{kN}^\pi] (p_{km}^\alpha + p_{km}^\beta) \right.$$

$$+ \tfrac{1}{2} [q_k^\alpha \ q_m^\alpha - (p_{km}^\alpha)^2 + q_k^\beta \ q_m^\beta - (p_{km}^\beta)^2] \ A_M^+$$

$$\left. + \tfrac{1}{2} (q_k^\alpha \ q_m^\beta + q_k^\beta \ q_m^\alpha) \ A_M^- \right\}) + \sum_k^{(M)} \sum_{l>k}^{(N)} \{ 2 \ (p_{kl}^\alpha + p_{kl}^\beta) [k_x \ l_x \ \beta_{kl}^\sigma$$

$$+ (k_y \ l_y + k_z \ l_z) \ \beta_{kl}^\pi] \} + \sum_m^{(M)} \sum_n^{(N)} \{ (p_{km}^\alpha \ p_{ln}^\alpha + p_{km}^\beta \ p_{ln}^\beta + p_{km}^\alpha \ p_{ln}^\beta$$

$$+ p_{km}^\beta \ p_{ln}^\alpha - p_{kn}^\alpha \ p_{lm}^\alpha - p_{kn}^\beta \ p_{lm}^\beta) [k_x \ l_x \ m_x \ n_x \ \gamma_{kl}^{\sigma\sigma} + (k_y \ l_y \ m_y \ n_y$$

$$+ k_z \ l_z \ m_z \ n_z) \ \gamma_{kl}^{\pi\pi} + k_x \ m_x \ (l_y \ n_y + l_z \ n_z) \ \gamma_{kl}^{\sigma\pi} + (k_y \ m_y + k_z \ m_z) \ l_x \ n_x \ \gamma_{kl}^{\pi\sigma}$$

$$+ (k_y \ m_y \ l_z \ n_z + k_z \ m_z \ l_y \ n_y) \ \gamma_{kl}^{\pi\pi*}] \}$$

IV. Applications

A. Ionization Potentials

One of the most accessible molecular properties available for testing the results of theoretical calculations are the molecular ionization potentials

$$A \longrightarrow A^+ + e \quad -\Delta H_f = I$$

The first ionization potential refers to the most weakly bound electron of the neutral molecule in the dilute gaseous phase, i.e. the energy liberated by removing an electron from the highest occupied orbital. Several experimental methods are available for measuring these for molecules and theoretical models can be set up so as to correlate them. Other ionization potentials may similarly be observed if, instead, an electron from an inner orbital is removed from the neutral molecule. Thus a molecule will have as many ionization potentials as occupied orbitals.

In contrast to the above process, the molecule, after losing its first electron, may lose a *second electron* and additional ones in successive steps. In such a case, however, the successive ionizations no longer refer to the neutral molecule. Very little is known about this latter process because the energy required to detach an electron from an already positively charged species is very high. In addition, the removal of more than one electron from a molecule often results in fragmentation.

The former ionizations are much better understood, and can now be measured experimentally by the recently developed technique of *photo-electron spectroscopy* [41]. A comparison between the experimental ionization potentials and the calculated orbital energies can then be made.

Unfortunately, the results are not as good as might have been expected. Whether the discrepancies originate from some inadequacy of the SCF calculation or from the invalidity of Koopman's theorem is, however, an open question. It is probably, to a first approximation only, that the removal of an electron from a molecular system can be considered as a "vertical" process which leaves the distribution and energies of the remaining electrons unaffected. Indeed, even the molecular geometry may be strongly affected by the process (Table 5).

Even though the differences are probably smaller in large molecules, they may still be large enough to explain why the calculated values of the ionization potentials are systematically higher than the observed ones. Such a behavior is compatible with a mechanism in which a redistribution of electronic densities accompanies the process of ionization, thus increasing the stability of the ion and lowering the ionization potential.

Table 5. *Variations in bond distances produced by the ionization of diatomic molecules* [42]

xy	r_{xy} (Å)	r_{xy}^+ (Å)
HH	0.74	1.06
ClCl	1.99	1.89
BH	1.23	1.21
CaH	2.00	1.73
CH	1.12	1.13
HCl	1.27	1.32
NN	1.09	1.12
OH	0.97	1.03

A possible check of the importance of this reorganization energy in the ions can be made by calculating their energy independently and estimating the ionization potential by the difference in heat of formation between the original compound and the ion. Although this procedure is certainly better than that based on Koopman's theorem, it is also costlier and still does not guarantee a complete agreement with experiment unless the geometry of the ionized species is known. Furthermore, only methods capable of handling open shells are suitable for such a purpose. Most attempts to correlate the ionization potentials have thus been made on the basis of Koopman's theorem.

In this respect, the CNDO and INDO methods have met with very little success, the values usually being too high by 3.5 eV to 5 eV. The error here seems too large even if provision is made for the possible reorganization of electrons in the molecule.

The best estimates have been obtained to date by using the MINDO and PNDO methods. In Tables 6 to 8 we show the ionization potential values obtained by each of these methods for alkanes and cycloalkanes, alkenes, acetylenes and aromatic compounds. Dewar and Klopman (PNDO) and Dewar et al. (MINDO/2) also compared their calculated inner orbital energies with experimental ionization potentials obtained from photoionization spectra. The ionization potentials of methane and ethane have also been calculated by the PNDO method along the more sophisticated procedure of minimizing separately the energy of the ion and that of the molecule. In these cases, the experimental value of the first ionization potential was reproduced accurately [43].

In Fig. 2 we compare, where possible, the values obtained by each method for a given compound. In general, it may be seen that the first ionization potential is better reproduced by the MINDO/2 method[b].

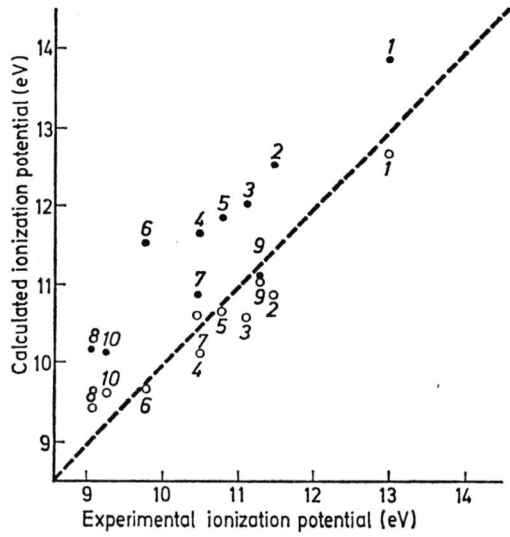

Fig. 2. Comparison between calculated and experimental ionization potentials. ---- Perfect correlation. ● Highest occupied orbital energies from PNDO; ○ Highest occupied orbital energies from MINDO/2. The numbers refer to the compounds in Tables 6 to 8

---

[b] Additional data on cyclic hydrocarbons has been published recently by Bodor, Dewar and Worley [59].

Table 6. *Comparisons of experimental ionization potentials with calculated orbital energies of alkanes and cycloalkanes*

Compound	Ionization potential(ev)	Orbital energy(ev) PNDO	Orbital energy(ev) MINDO/2
1 Methane	12.99	13.88	12.69
2 Ethane	11.49	12.51	10.87
		13.04	
	14.74	14.98	
	19.18	20.80	
3 Propane	11.07	12.01	10.56
		12.49	
		12.85	
	13.17	13.73	
		13.90	
	15.17	14.68	
	15.70	15.46	
	18.57	19.67	
Cyclopropane	10.06	—	10.27
4 *n*-Butane	10.50	11.63	10.13
		12.39	
	12.36	12.78	
		13.07	
		13.21	
		14.13	
	14.13	14.36	
		14.47	
	15.69	15.79	
5 *i*-Butane	10.78	11.88	10.63
		12.54	
	12.54	13.48	
		13.79	
	14.51	14.59	
		15.40	
	18.63	18.68	
Cyclobutane	—		9.80
Cyclopentane	10.49	—	9.76
6 Cyclohexane	9.79	11.51	9.65
	11.33	12.23	
		12.59	
	12.22	12.69	
		13.48	
	14.37	15.10	
		15.51	

Table 7. *Comparison of experimental ionization potentials with calculated orbital energies of alkenes (dienes)*

Compound	Ionization Potential(ev)	Orbital energy(ev) PNDO	Orbital energy(ev) MINDO/2
7 Ethylene	10.48	10.86	10.60
		12.76	
	12.50	12.94	
	14.39	15.25	
	15.63		
	19.13	19.18	
1-Butene	9.59	—	9.74
cis-2-Butene	9.12	—	9.32
trans-2-Butene	9.12		9.34
iso-Butene	9.17		9.36
8 trans-1.3-Butadiene	9.08	10.16	9.40
	11.25	11.70	
		11.83	
	12.14	12.58	
		13.09	
	12.23	14.39	
		14.71	
	18.78	17.99	
		19.24	
Allene	9.83		9.74

Table 8. *Comparison of experimental ionization potentials with calculated orbital energies of acetylenes and aromatic compounds*

Compound	Ionization potential(ev)	Orbital energies(ev) PNDO	Orbital energies MINDO/2
9 Acetylene	11.36	11.06	11.02
	16.27	13.63	
	18.33	18.08	
Diacetylene	10.17		9.80
10 Benzene	9.25	10.15	9.01
	11.49	11.54	
	12.19	12.72	
		12.86	
	13.67	13.45	
	14.44	15.67	
	16.73	16.07	
	18.75	18.98	
Toluene	8.82	—	9.18
Azulene	7.43	—	7.95

B. Heats of Formation

Although the CNDO and INDO methods in their original form do not give anything close to the experimental heats of formation, there have been attempts to reparametrize them so as to allow some correlation to be found. Apparently, however, the best that could be achieved so far in this respect was Wiberg's [30] modified CNDO/2, which led to heats of formation proportional to the experimentally observed ones.

Dewar and Klopman (PNDO) and Dewar et al. (MINDO/1) have calculated the heats of formation of a large number of *hydrocarbons* with good accuracy. These calculations all require the use of an artificially lowered nuclear-nuclear repulsion term. Such an approximation usually results also in unrealistically short values for the bond distances. The reason why such a correction has to be introduced in order to obtain good heats of formation is still unknown and is under investigation. The net result is the painful dilemma of having to choose between a method that gives good heats of formation and poor bond distances, or poor heats of formation and good bond distances. This problem, which is also present in ab initio calculations, may be related to the neglect of the changes in correlation energy when a molecule is formed. Its solution, if it can be found, would probably be the most important contribution in the field of semi-empirical calculation of large molecules.

Both PNDO and MINDO/1 thus use standard values of bond distances (Table 9). In the MINDO/2 program however, Dewar and Haselbach [35] seem to have solved the problem outlined above. Their calculation provide both good heats of formation and bond distance for *Hydrocarbons and Nitrogen and Oxygen heteromolecules*, thus opening a new dimension for application of their method.

Table 9. *Standard bond distances for C—C and C—H bonds in PNDO and MINDO methods*

Bond	Hybridization	Bond Length
C—C	sp^3-sp^3	1.534 Å
	sp^3-sp^2	1.520
	sp^3-sp	1.459
	sp^2-sp^2	1.483
C≂C	(Aromatic)	1.397
C=C		1.337
C≡C		1.205
C—H	sp^3	1.093
	sp^2	1.083
	(Aromatic)	1.084
	sp	1.059

Table 10. *Comparison of calculated and observed heats of formation of alkanes and cycloalkanes*[a]

Compound	ΔH_f obs.	PNDO ΔH_f cal.	PNDO $\delta\Delta H_f$	MINDO/1 ΔH_f cal.	MINDO/1 $\delta\Delta H_f$	MINDO/2 ΔH_f cal.	MINDO/2 $\delta\Delta H_f$
Methane	−17.9	−17.9	0.0	−17.9	0.0	−11.9	+6.0
Ethane	−20.2	−22.9	−1.9	−20.2	0.0	−21.9	−1.7
Ethane (eclipsed)	−17.3	−22.1	−4.8	—	—	−19.8	−2.5
Propane	−24.8	−26.8	−2.0	−24.9	−0.1	−24.9	−0.1
n-Butane	−30.2	−31.5	−1.3	−29.4	+0.7	−28.4	+1.8
iso-Butane	−32.1	−32.24	−0.1	−31.8	+0.3	—	—
2-Methylbutane	−32.2	−31.9	+0.3	−31.8	+0.4	—	—
n-Pentane	−35.0	−36.2	−1.2	−34.0	+1.0	—	—
iso-Pentane	−36.9	36.6	+0.3	37.1	−0.2	—	—
2-Methylpentane	−36.9	−35.5	+1.4	−37.1	−0.2	—	—
neo-Pentane	−39.7	−37.6	+2.3	−40.6	−0.9	—	—
Cyclopropane	+12.7	—	—	+16.4	+3.7	+ 3.8	+8.9
Methylcyclopropane	+25.5	—	—	−26.2	−0.7	—	—
Cyclobutane	+ 6.4	+ 6.4	0.0	+11.3	+4.9	—	—
Methylcyclobutane	(− 0.8)	—	—	+ 4.5	(+5.3)	—	—
Cyclopentane	+12.7	+12.1	−0.6	+16.4	+3.7	—	—
Methylcyclopentane	−25.5	—	—	−26.19	−0.7	—	—
Cyclohexane (chair)	−29.4	−29.6	−0.2	—	—	−26.6	+2.8
Cyclohexane (boat)	−24.1	−24.4	−0.3	—	—	−22.0	+2.1
Methylcyclohexane	−37.0	−35.3	+1.7	−34.9	+2.1	—	—
Dicyclopropyl	+31.0	+36.1	+5.1	+39.5	+8.4	—	—
Spiropentane	+44.2	+46.3	+2.1	+44.2	−2.0	—	—
1,3-Dimethylbicyclo[1.1.0]butane	+39.7	+40.5	+0.8	+31.0	−8.7	—	—
Cubane	+148.7	+116.9	−31.8	+116.9	−31.8	—	—

[a] Values taken from (i) Molecular Orbital Theory of Organic Chemistry. M. J. S. Dewar, p. 459. New York: McGraw-Hill Book Company 1969; (ii) Dewar, M. J. S., Klopman, G.: J. Am. Chem. Soc. 89, 3089 (1967); (iii) Baird, N. C., Dewar, M. J. S.: J. Am. Chem. Soc. 89, 3966 (1967); (iv) Baird, N. C., Dewar, M. J. S.: J. Chem. Phys. 50, 1262 (1969); (v) Dewar, M. J. S., Haselbach, E.: (to be published).
[b] Estimated value C. T. Mortimer, Reaction Heats and Bond Strengths. New York: Pergamon Press, Inc. 1962.

Table 11. *Comparison of calculated and observed heats of formation of alkenes and cycloalkenes*

Compound	ΔH_f obs.	PNDO		MINDO/1		MINDO/2	
		ΔH_f cal.	$\delta\Delta H_f$	ΔH_f cal.	$\delta\Delta H_f$	ΔH_f cal.	$\delta\Delta H_f$
Ethylene	+12.5	+12.5	0.0	+13.8	+1.3	+16.4	
Propene	+ 4.9	+ 4.5	−0.4	+ 5.2	+0.3		
1-Butene	0.0	− 0.4	−0.4	+ 0.8	+0.8	+ 1.3	
cis-2-Butene	− 1.7	+ 0.3	+2.0	−	−	− 2.2	
trans-2-Butene	− 2.7	− 2.9	−0.2	− 3.0	−0.3	− 6.2	
iso-Butene	− 4.0	− 3.8	+0.2	− 5.9	−1.9	0.4	
s-cis-1,3-Butadiene	+28.5	+27.2	−1.3	+25.7	−0.6	+29.6	
s-trans-1,3-Butadiene	+26.3	+26.1	−0.2	+47.14			
Methylenecyclopropane	+48.0			+67.61			
2-Methylmethylenecyclopropane	+39.4	+40.2	+0.8	+78.54			
Ethylidenecyclopropane	+36.1	+37.6	+1.5				
Dimethylenecyclopropane		—					
Trimethylenecyclopropane		—					
Cyclopropene	+66.6	+58.7	−7.9	+66.6	0.0		
1-Methylcyclopropene	+58.2			+54.81			
3-Methylcyclopropene				59.40			
1,2-Dimethylcyclopropene	+46.4	+42.4	−4.0	+43.4	−3.0		
Methylenecyclopropene				+75.27			
Cyclobutene	+35.0	+35.1	0.0	+40.9	+5.9		
1-Methylcyclobutene	(+27.7)			+30.17			
1,2-Dimethylcyclobutene	+22.4	+18.0	−4.4	+19.6	−3.8		
Spiropentadiene							
Methylenecyclopentane	+ 1.90			+ 1.27			
Cyclopentene	+ 7.7	+ 3.3	−4.4	+ 8.1	+0.4		
1-Methylcyclopentene	− 2.0			− 2.1			
Cyclopentadiene	+32.4	+27.8	−4.6	+30.2	−2.2		
Fulvene				+39.93			
6,6-Dimethylfulvene	+31.0	+27.6	−3.4	+19.6	−11.4		
Cyclohexene	− 1.7	+ 3.2	+4.9	+ 3.6	−4.9		
Allene	+45.9	29.9				+40.0	

Table 12. *Comparison of calculated and observed heats of formation of acetylenes and aromatic compounds*

Compound	ΔH_f obs.	PNDO		MINDO/1		MINDO/2	
		ΔH_f cal.	$\delta \Delta H_f$	ΔH_f cal.	$\delta \Delta H_f$	ΔH_f cal.	$\delta \Delta H_f$
(Acetylenes)							
Acetylene	+54.3	+32.0				+53.4	
Methylacetylene	+44.29	+16.69				+53.4	
Diacetylene	+111.8					+101.8	
(Aromatics)							
Benzene	+19.8	+19.9	+0.1	+20.2	+0.4	+20.2	
Toluene	+12.0	+11.7	−0.3	+10.2	−1.8	+11.1	
o-Xylene	+ 4.5	+ 4.3	−0.2	− 0.3	−4.8		
m-Xylene	+ 4.1	+ 3.6	−0.5	+ 0.2	−3.9		
p-Xylene	+ 4.3	+ 3.5	−0.8	0.9	−4.3		
Styrene	+35.2	+34.3	−0.9	+32.9	−2.3		
Napthalene	+36.3	+37.9	+1.6	+27.5	+8.8		
Azulene	+68.9	+72.1	+3.2				

In Tables 10 to 12 we show the heats of formation calculated by the various methods, together with their deviation from the experimentally observed values for alkanes and cycloalkanes, alkenes and cycloalkenes, and acetylenes and aromatic compounds. Table 13 shows a comparison of heats of formation of hydrocarbon radicals calculated by the MINDO methods. Finally, in Tables 14 and 15 we show the results of MINDO/1 calculations on a selection of oxygen- and nitrogen-containing compounds.

Table 13. *Heats of formation of hydrocarbon radicals calculated by MINDO/1 and MINDO/2*

Radical	ΔH_f obs.	MINDO/1		MINDO/2	
		ΔH_f cal.	$\delta \Delta H_f$	ΔH_f cal.	$\delta \Delta H_f$
$CH_3 \cdot$	+ 34.0	+33.5	−0.5	30.2	
$CH_3CH_3 \cdot$	+ 25.7	+28.6	+2.9		
$CH_3CH_2CH_2 \cdot$	+ 21.0	+23.9	+2.9		
$(CH_3)_2CH \cdot$	+ 17.6	+19.7	+2.1		
$CH_3CH_2CH_2CH_2 \cdot$	+ 17.0	+19.3	+2.3		
$CH_3CH_2CH \cdot CH_3$	+ 12.4	+15.2	+2.8		
$(CH_3)_3C \cdot$	+ 6.8	+ 7.4	+0.6		
Allyl	+ 37.0	+46.0	+9.0	35.2	
$C_6H_5 \cdot$	71.0			71.7	
	80.0				
$C_6H_5CH_2 \cdot$	45.0			48.6	
CH_4^+	282.6			272.4	
$C_6H_6^+$	233.3			243.9	

Table 14. *Heats of formation of oxygen containing compounds*

Compound	ΔH_f obs.	MINDO/1	
		ΔH_f cal.	$\delta \Delta H_f$
Ethylene oxide	−12.19	− 0.49	−12
Furan	− 8.29	− 7.27	− 1
Phenol	−23.05	−26.54	+ 3
Anisol	−19.00	−25.86	+ 7
Benzyl alcohol	−22.39	−25.27	+ 3
o-Cresol	−30.74	−36.87	+ 6
m-Cresol	−31.63	−36.64	+ 5
p-Cresol	−29.97	−36.61	+ 7

Table 15. *Heats of formation of nitrogen containing compounds*

Compound	ΔH_f obs.	MINDO/1	
		ΔH_f cal.	$\delta \Delta H_f$
Ammonia	−11.04	−11.03	0
Methylamine	− 6.70	− 6.70	0
Dimethylamine	− 6.60	− 3.67	− 3
Trimethylamine	−10.55	− 1.78	− 9
Ethylamine	−11.60	− 7.05	− 5
n-Butylamine	−22.50	−22.63	0
s-Butylamine	−25.20	−25.58	0
t-Butylamine	−28.65	−25.85	− 3
Aniline	+20.80	+14.85	+ 6
Hydrazine	+22.75	+22.73	0
Methylhydrazine	+23.35	+20.60	+ 3
Pyrrole	+24.61	+19.48	+ 5
Pyridine	+33.61	+33.65	0
2-Methylpyridine	+23.65	+22.91	+ 1
Pyrazine	+46.86	+48.11	− 1
Pyrimidine	+46.99	+32.32	+15

In recent months, increasing interest has been shown in the theoretical calculation of the stability of carbonium ions [c]. Table 16 shows the results of such calculations performed by the PNDO method [44].

Table 16. *Heats of formation of carbonium ions from gaseous carbon and hydrogen atoms*

Compound	ΔH_f (eV)	
	Observed	Calculated
CH_3^+	− 3.04	− 2.80
CH_5^+ (trigonal bipyramid)		− 9.10
CH_5^+ (sp^3 hybrids)		− 9.46
$C_2H_5^+$	−16.5	−16.2
$C_2H_7^+$		−22.03
$C_5H_{13}^+$		−57.89
Norbornane cation (classical)		−66.22
Face protonated nortricyclene		−66.34
Edge protonated nortricyclene		−68.16
Corner protonated nortricyclene		−68.02

---

[c] Additional papers dealing with this subject have appeared in the literature after completion of this work [58,60].

The interest in this area stems from attempts to assess the relative stability of various possible structures of carbonium ions. Therefore absolute values for the heats of formation are not necessarily required. CNDO calculations can thus be used equally well to determine the relative stability of isomers. Such calculations have been performed by Wiberg [45] and are illustrated in Table 17.

Table 17. *Energies of formation of some cations* [45] $RH \rightarrow R^+ + H \cdot + e^-$

R^+	ΔE_{eV} (Calc.)
CH_3^+	25.00
$CH_3CH_2^+$	22.39
$CH_3^+CHCH_3$	20.57
$t\text{-}C_4H_9^+$	19.06
$i\text{-}C_4H_9^+$	21.89
$\triangleright\!-CH_2{}^+$	22.25
$\diamondsuit\!-CH_2{}^+$	21.81
$\diamondsuit{}^+$	20.82

C. Dipole Moments

The dipole moment of a compound is a function of the distribution of charge within the molecule, and appears to be a sensitive test for the accuracy of the compound's molecular wave functions. The dipole moment of a molecule can be approximated for a given direction as the sum of two components, μ_Q, the contribution from net charge densities on the atoms, and for each atom A, μ_{sp} (A), an atomic polarization moment produced by the distortion of the electronic cloud around the atom. The atomic polarization moment results essentially from the mixture of s and p orbitals and, for a heteroatom, includes μ_{lp}, the lone pair moment.

Thus, the x component of the dipole moment for a given molecule is

$$\mu_{total}^x = \mu_Q^x + \sum_{A'} \mu_{sp}^x (A') \tag{82}$$

where the summation is over all non-hydrogen atoms (i.e. the ls atomic orbital only is used in the basis set of hydrogen, and hence no sp mixing can be taken into account, and

and

$$\mu_Q^x = 2.5416 \sum_A^{\text{all atoms}} (Z_A - P_{AA})\, x_A \tag{83}$$

$$\mu_{sp}^x (A) = -14.674 \left(\frac{P_{2s(A)\, 2px(A)}}{Z_A'} \right) \tag{84}$$

Here Z_A is the core charge on A (e.g. the nuclear charge less the number of inner core electrons), P_{AA} is the total charge on atom A, x_A is the cartesian coordinate of atom A, and Z_A' is the Slater orbital exponent for $2\,s$ and $2\,p$ orbitals of atom A.

Pople and Segal [46] (CNDO/1) have calculated the dipole moments for a number of diatomic molecules with limited success. Improved results have, however, been obtained by the CNDO/2 approximation. Thus Pople and Segal [26] (CNDO/2), working with a large number of AB_2 and AB_3 molecules, and Pople and Gordon [47] (CNDO/2), working with a large number of organic compounds containing *nitrogen, oxygen, and fluorine*, have obtained good agreement.

Segal and Klein [48] (CNDO/2), working with some small molecules, have shown in cases where no ambiguity can arise due to the cancellation of large terms, that both the magnitude and sign of derivatives are quite well reproduced. Bloor et al. [49] (CNDO/2), working with a selection of nitrogen- and oxygen-containing heterocycles and some fluorine-containing compounds, and Davies [50] (CNDO/2), working with fluorinated benzenes and nitrobenzenes and the radical anions of nitrobenzene have reproduced dipole moments well. The dipole moments of a few small compounds calculated by Yonezawa et al. [32a] (CNDO/2 level) give values in excess of the experimental figure by about 1 Debye.

At the INDO level Pople, Beveridge and Dobosh [27] have compared the dipole moments calculated for a selection of AB_2 and AB_3 molecules with CNDO/2 results. In general the values are not too dissimilar.

Yonezawa et al. [32b] have calculated the dipole moment of cis-1,3-butadiene together with three oxygen-containing compounds. In only one case, trans-acrolein, is it possible to make a comparison with an experimental value, and here the calculated value is 0.45 debyes greater than the observed one.

Dewar et al. [35] (MINDO/2) have calculated the dipole moment of an AB_2 and an AB_3 type molecule. In both cases the values are larger than those of CNDO/2 or INDO. This trend is also apparent when MINDO/1 is applied to a selection of oxygen- and nitrogen-containing compounds. As yet no extensive dipole moment calculations have been reported using the MINDO/2 method.

Finally, PNDO has been used to calculate the dipole moments of some saturated hydrocarbons with reasonable agreement with experiment. In Table 18 we show the values obtained by each method for small molecules of the general form AB, AB_2 and AB_3. In Tables 19 to 23 we show the values obtained by each method for hydrocarbons, fluorine-containing compounds, oxygen-containing compounds, nitrogen-containing compounds and other miscellaneous compounds. In Fig. 3 and 4 we compare, where possible, these values, to the experimentally observed ones.

Table 18. *Calculated and experimental dipole moments for AB, AB_2, and AB_3 molecules*

Molecule	Obs.	Dipole moments[a] (debyes)		
		Cal. CNDO/2	Cal. INDO	Cal. MINDO/1
(AB)				
NO	0.16	−0.16		
CO	0.11	−1.00		
HF	−0.13[e]	−1.03[e]		
	1.82	1.85[d]		
(AB_2)				
BeH_2	—	0	0	
BH_2 ($^2A_1$)	—	0.51	0.32	
CH_2 ($^1A_1$)	—	2.26	2.17	
CH_2 ($^3B_1$)	—	0.75	0.53	
NH_2 ($^2B_1$)	—	2.16	2.12	
NH_2 (^2II $^2A_1$)	—	0.87	0.79	
OH_2	1.8	2.08	2.14	2.79[c]
FH_2	—	0	0	
BO_2	—	0	0	
CO_2	—	0	0	
BeF_2	—	0	0	
NO_2	±0.4	−0.75	−0.79	
BF_2	—	0.05	−0.29	
O_3	±0.58	−1.26	−1.09	
CF_2	—	0.53	0.26	
NF_2	—	−0.12	−0.38	
OF_2	±0.297	−0.21	−0.40	

Table 18 (continued)

Molecule	Obs.	Dipole moments[a] (debyes)		
		Cal. CNDO/2	Cal. INDO	Cal. MINDO/1
(AB_3)				
BH_3	—	0	0	
CH_3	—	0	0	
NH_3	1.47	−2.08	1.90	2.13[c]
H_3O	—	0	0	
BF_3	—	0	0	
CF_3	—	−0.17	−0.68	
NF_3	±0.23	0.05	−0.48	
HCN	2.42	2.48[d]		
	2.95[e]	4.34[e]		

[a] The convention of a positive sign meaning: the atom furthest to the right in the formula is at the negative end of the dipole, is employed. Unless otherwise stated, values are taken from Pople, J. A., Beveridge, D. L., Dobosh, P. A., reference [27].

[b] Values taken from Segal, G. A., Klein, M. L., reference [48].

[c] Values taken from Baird, N. C., Dewar, M. J. S., Sustmann, R.: J. Chem. Phys. *50*, 1275 (1969).

[d] Values taken from Bloor, J. E., Gilson, B. R., Billingsley II, F. P.: Theoret. Chim. Acta (Berl.) *12*, 360 (1968).

[e] Yonezawa, T., Yamaguchi, K., Kato, H., reference [32a].

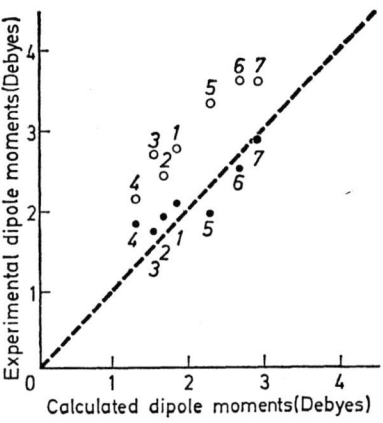

Fig. 3. Comparison between calculated and observed dipole moments. – – – – Perfect correlation. ● Calculated dipole movements CNDO/2. ○ Calculated dipole movements MINDO/I. The numbers refer to compounds in Table 21

Table 19. *Dipole moments (debyes) of hydrocarbons*[a])

Compound	Obs.	Cal. CNDO/2	Cal. INDO level	Cal. PNDO
Hydrocarbons				
Propane	0.083	0.00 0.03[d])		0.03[b])
Propene	0.364 0.350[d])	0.36 0.50[d])		
Propyne	0.75	0.43 0.56[d])		0.24[b])
2-Methylpropane	0.132	0.00		
2-Methylpropene	0.503	0.65		
2-Methyl-1,3-butadiene	0.292	0.25		
Isobutane	0.13			0 05[b])
cis-2-Butene	—			0.08[b])
cis-1,3-Butadiene	—		0.31[e])	0.04[b])
Butenyne	—	0.37[d]) 0.17[d])		
Toluene	0.43 0.31[c])	0.21 0.22[c])		

[a]) Unless otherwise stated, values are taken from Pople, J. A., Gordon, M., reference [47]).
[b]) Dewar, M. J. S., Klopman, G., reference [28]).
[c]) Bloor, J. E., Breen, D. L.: J. Phys. Chem. *72*, 716 (1968).
[d]) Fischer, H., Kollmar, H.: Theoret. Chim. Acta (Berl.) *13*, 213 (1969).
[e]) Kato, H., Konishi, H., Yamabe, H., Yonezawa, T.: Bull. Chem. Soc. Jap. *40*, 2761 (1967).

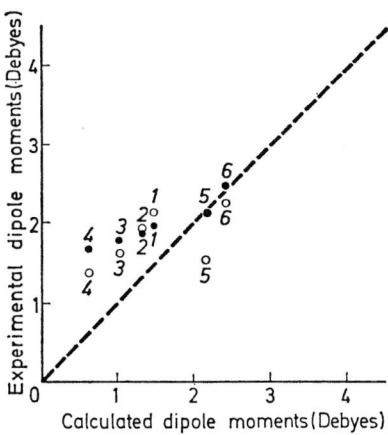

Fig. 4. Comparison between calculated and experimental dipole moments. — — — Perfect correlation. ● Calculated dipole movements CNDO/2. ○ Calculated dipole movements MINDO/1. The numbers refer to compounds in Table 22

Table 20. *Dipole moments (debyes) of fluorine compounds*

Compound	Obs.	Cal. CNDO/2
Hydrogen fluoride	1.8195	1.85
Methyl fluoride	1.855	1.66
Methylene fluoride	1.96	1.90
Fluoroform	1.645	1.66
Ethyl fluoride	1.96	1.83
1,1-Difluoroethane	2.30	2.23
1,1,1-Trifluoroethane	2.32	2.18
Fluoroethylene	1.427	1.51
1,1-Difluoroethylene	1.37	1.02
cis-1,2-Difluoroethylene	2.42	2.83
Fluoroacetylene	0.75	1.04
n-Propyl fluoride (trans)	2.05	1.84
trans-1-Fluoropropene	1.85	1.67
cis-1-Fluoropropene	1.46	1.59
2-Fluoropropene	1.60	1.69
3-Fluoropropene (s-cis)	1.765	1.83
3,3,3-Trifluoropropene	2.45	2.34
3,3,3-Trifluoropropyne	2.36	2.48
2-Fluoro-1,3-butadiene	1.417	1.65
Fluorobenzene	1.66	1.66
Trifluoromethylbenzene	2.86	2.73[c]
o-Difluorobenzene	2.40	2.88[a]
m-Difluorobenzene	1.58	1.65[a]
p-Difluorobenzene	0.0	0.0[a]
1,2,3-trifluorobenzene	—	3.33[a]
1,2,4-trifluorobenzene	—	1.66[a]
1,2,3,4,5-pentafluorobenzene	—	1.70[a]
Hexafluorobenzene	0.0	0.0[a]
Formyl fluoride	2.02	1.98[b]

[a] Davies, D. W., reference [50].
[b] Dewar, M. J. S., Klopman, G., reference [28].
[c] Bloor, J. E., Breen, D. L.: J. Phys. Chem. *72*, 716 (1968).

Table 21. *Dipole moments of oxygen-containing compounds*

Compound	Obs.	Cal. CNDO/2	Cal. INDO level	Cal. MINDO/1[d]
1 Water	1.846	2.10 2.14[c] 2.92[e]		2.79
2 Methanol	1.69	1.94		2.48
Ethanol	1,70			2.35
Propanol	1.64			2.21
Allyl alcohol	1.63			2.27
3 Phenol	1.55	1.73 1.76[c]		2.72
Benzyl alcohol	1.71			2,18
4 Dimethyl ether	1.30	1.83		2.17
Diethyl ether	1.18			2.03
Ethylene oxide	1.88			3.19
Furan	0.66[d] 0.72[b]	0.80[b]		1.42
Anisole	1.35			2.38
5 Formaldehyde	2.339 2.17[c] 2.30[e]	1.98 1.92[c] 3.05[e]		3,35
6 Acetaldehyde	2.68	2.53		3.64
Propionaldehyde	2.52	2.46		—
Acetylacetylene	2.40	2.85		—
7 Acetone	2.90	2.90		3.63
Acroline (*s-trans*)	3.11	2.63		
Methyl vinyl ketone	3.16	2.92		
Ketene	1.414	1.30		
Methyl Ketene	1.79	1.35		
Formic acid	1.415	0.87 1.34[a]		
Benzaldehyde	2.72	2.50[c]		
trans-Acrolein	3.11[f]		3.56[f]	
cis-Acrolein	—		3.66[f]	
cis-Glyoxal	—		5.10[f]	

[a] Bloor, J. E., Gilson, B. R., Billingsley II, F. P., Theoret. Chim. Acta (Berl.) *12*, 360 (1968).

[b] Bloor, J. E., Breen, D. L., reference [49].

[c] Bloor, J. E., Breen, D. L.: J. Phys. Chem. *72*, 716 (1968).

[d] Baird, N. C., Dewar, M. J. S., Sustmann, R.: J. Chem. Phys. *50*, 1275 (1969).

[e] Yonezawa, T., Yamaguchi, K., Kato, H., reference [32a].

[f] Kato, H., Konishi, H., Yamabe, H., Yonezawa, T.: Bull. Chem. Soc. Jap. *40*, 2761 (1967).

Table 22. *Dipole moments (debyes) of nitrogen-containing compounds*

Compound	Obs.	Cal. CNDO/2	Cal. MINDO/1[d]
1 Ammonia	1.468	1.97 2.09[a] 3.11[e]	2.13
2 Methylamine	1.326	1.86	1.88
Dimethylamine	1.03	1.76	1.63
4 Trimethylamine	0.612	1.68	1.39
Ethylamine	1.22	—	1.87
n-Butylamine	1.00		1.82
Aniline	1.48[d] 1.53[c]	1.53[c]	1.97
Methylhydrazine	1.68		0.36
Pyrrole	1.84[d] 2.20[b]	−2.00[b]	1.28
5 Pyridine	2.20	2.19[b]	1.54
Pyridazine	3.97	3.76[b]	
6 Pyrimidine	2.42	2.46[b]	2.27
Quinoline	2.31	2.34[b]	
Isoquinoline	2.75	2.20[b]	
Indole	2.00	1.86[b]	
Imidazole	4.02	−4.09[b]	
Pyrazole	2.21	2.71[b]	
1,2,3-Triazole	1.79	4.30	
1,2,5-Triazole		0.20	
1,2,3,5-Tetrazole	5.15	2,35	
1,2,3,4-Tetrazole		5.23	
Cyanobenzene	3.93	3.23[c]	

[a] Bloor, J. E., Gilson, B. R., Billingsley II, F. P.: Theoret. Chim. Acta (Berl.) *12*, 360 (1968).
[b] Bloor, J. E., Breen, D. L., reference [49].
[c] Bloor, J. E., Breen, D. L.: J. Phys. Chem. *72*, 716 (1968).
[d] Baird, N. C., Dewar, M. J. S., Sustmann, R.: J. Chem. Phys. *50*, 1275 (1969).
[e] Yonezawa, T., Yymaguchi, K., Kato, H., reference [32a].

Table 23. *Dipole moments of miscellaneous compounds*

Compound	Obs.	Cal. CNDO/2
Nitrogen trifluoride	0.235	0.43
Difluoramine	1.93	2.13
Nitrous acid	1.85	2.27
Nitric acid	2.16	2.24
Cyano fluoride	1.68	1.55
Formyl fluoride	2.02	2.16
Carbonyl fluoride	0.951	1.42
Acetyl fluoride	2.96	2.84
Acetyl cyanide	3.45	2.80
Isocyanic acid	1.59	1.88
Methyl isocyanate	2.81	1.80
Formamide	3.71	3.79
Nitromethane	3.46	4.38
Nitrobenzene	4.28 4.40[c]	5.33 4.95[d]
Isoxazole	3.01	3.17[c]
Oxazole	1.40	1.34[c]
1,2,5-Oxadiazole	3.36	−3.52[c]
1,3,4-Oxadiazole	3.0	2.89[c]
Sydnone	7.31	6.82[b]
o-Fluoronitrobenzene	—	6.28[a]
m-Fluoronitrobenzene	—	4.66[a]
p-Fluoronitrobenzene	2.87	3.71[a]

[a] Davies, D. W., reference [50].
[b] Bloor, J. E., Gilson, B. R., Billingsley II, F. P.: Theoret. Chim. Acta (Berl.) *12*, 360 (1968).
[c] Bloor, J. E., Breen, D. L., reference [49].
[d] Bloor, J. E., Breen, D. L.: J. Phys. Chem. *72*, 716 (1968).

D. Molecular Geometries and Force Constants

With the exception of the PNDO and MINDO/1 methods, molecular geometries and force constants have been calculated by all the available semi-empirical methods. Pople and Segal found that CNDO/1 was able to reproduce bending force constants but not bond lengths. This was corrected by additional approximations leading to CNDO/2. This latter method satisfactorily reproduced bond angles (Table 24), bond lengths (Table 25) and bending force constants (Table 26) for a large number of molecules. However, the CNDO/2 calculated stretching force constants remain too large.

Kroto and Santy [51] have used the CNDO/2 method to calculate bond angles of a few molecules in their excited states. Again their results seem good. Subsequently [52], they performed a far more laborious open-shell calculation which resulted in only slightly improved bond angles.

Wyberg, in his modified CNDO/2 method [30], calculated bond angles which are in good agreement with experiment. Del Bene and Jaffe [31] in their modified CNDO/2 method, however, were unable to reproduce satisfactory bond angles.

At the INDO level, the same excellent agreement with experiment has been obtained by Pople et al. (INDO) and by Dixon [33] in his EMDZO method as in the original CNDO/2 (Table 24).

The recently reparametrized MINDO method (MINDO/2) [35], seems to provide equally good values for bond distances and force constants.

Tables 24 to 26 show comparisons of experimental and calculated data for the various methods.

E. Ultraviolet Spectra

One of the traditional tests, and the main success, of the quantum mechanical calculations on conjugated molecules was the prediction of UV transitions. Attempts have also been made to use all-valence electron calculations for this purpose. Early attempts, however, met with relative failure. Thus the UV spectra of methyl-substituted borazines and benzenes were calculated by Kuznetsof and Shriver [53] but did not correlate very well with the experimentally observed ones. Clark et al [29b] using virtual orbitals obtained by the CNDO/2 method, encountered similar difficulties.

The calculation of the geometries of species in their excited states seems, however, to be more successful. Thus satisfactory bond angles in a few selected molecules were calculated by Santry and Kroto [51]. More recently Del Bene and Jaffe reparametrized the CNDO method and calibrated it to reproduce the spectra of benzene and pyridine. Their

Table 24. *Comparison of calculated and experimental equilibrium bond angles*

Compound	Angle (BAB)	Exptl. Bond Angle	Cal. CNDO/2	Cal. INDO	Cal. EMZDO	Cal. CNDO/2 (Wyberg)
BeH$_2$	—	—	180.0	180.0		
BH$_2$ ($^2A_1$)		—	136.6	130.0		
BH$_2$ ($^2II - ^2B_1$)		—	180.0	180.0		
CH$_2$ ($^1A_1$)		103.2	108.6	107.2	105.8	
CH$_2$ ($^3B_1$)		180	141.4	132.4	131.0	
NH$_2$ ($^2B_1$)		103.3	107.3	107.2	106.0	
NH$_2$ ($^2II - ^2A_1$)		144	145.1	140.3	141.2	
OH$_2^+$ ($^2B_1$)		—	118.7	123.4		
OH$_2$		104.45	107.1	108.6	106.6	
FH$_2$		—	180.0	180.0		
BO$_2$		180	180.0	180.0		
CO$_2^+$		180	180.0	180.0		
CO$_2$		180	180.0	180.0		
BeF$_2$		180	180.0	180.0		
NO$_2^+$		180	180.0	180.0		
CO$_2^-$		134	142.3	140.8		
NO$_2$		132	137.7	138.5		
BF$_2$		—	124.6	122.9		
NO$_2^-$		115.4	118.3	118.6		
O$_2$		116.8	114.0	115.4		
CF$_2$		(100 or 108°)	104.6	103.6		
NF$_2$		104.2	102.5	101.7		
OF$_2$		103.8	99.2	99.0		
BH$_3$		—	120.0	120.0		
CH$_3$		(12.0)	120.0	120.0		
OH$_3^+$		117	113.9	120.0		
NH$_3$		106.6	106.7	109.7	108.1	
H$_2$O		—	120.0	120.0		
CO$_3^-$		120	120.0	120.0		
BF$_3$		120	120.0	120.0		
NO$_3^-$		120	120.0	120.0		
CF$_3$		111.1	113.5	111.6		
NF$_3$		102.5	104.0	101.0		
Ethane	(H—C—C)	110.5				111.5
Propane	(C—C—C)	112.4				113.3
	(H—C—H)	106.1				105.6
Ethylene	(C—C—H)	122.3				123.3
Allene	(C—C—H)	120.2				122.8
						121.5

Table 25. *Comparison of experimental and calculated bond lengths*

Compound	Bond	Experimental Bond Length	cal[d]) CNDO/2	cal[e]) MINDO/2
Hydrogen (molecular)		0.742 Å	0.746 Å	
Hydrogen fluoride		0.917	1.004	
OH		0.971	1.026	
NH		1.038	1.062	
Nitrogen (molecular)		1.094	1.140	
N_2^+		1.116	1.113	
CH		1.120	1.109	
O_2^+		1.123	1.095	
Carbon monoxide		1.128	1.190	
NO		1.151	1.152	
Oxygen (molecular)		1.207	1.132	
BH		1.233	1.193	
BeH		1.343	1.324	
Lithium hydride		1.595	1.568	
Methane	C—H	1.091[b])		1.196
Ethane		1.091[a])	1.117	1.103
Ethane (ecl)		—	—	1.103
Ethylene		1.086[a])	1.110	1.093
Acetylene		1.058[a])	1.093	1.069
Diacetylene		1.046[b])	—	1.069
Allene		(1.06,1.081)[b])	—	1.093
Cyclopropane		1.089[b])	—	1.103
Cyclobutane		1.092[b])	—	1.103
Hydrogen cyanide		1.065[a])	1.093	—
Formaldehyde		1.102[a])	1.116	—
Ethane	C—C	1.536[a])	1.476	1.524
Ethane (ecl)		—	—	1.524
Propane		1.50[b])	—	1.534
n-Butane		1.51[b])	—	1.540
i-Butane		1.540[b])	—	1.534
Cyclopropane		(1.51,1.524)[b])	—	1.519
Cyclobutane		(1.548,1.567)[b])	—	1.550
Cyclopentane		1.52[b])		
Cyclohexane (chair)		(1.53,1.540)[b])	—	1.549
Cyclohexane (boat)		—	—	1.551
Benzene		1.393[b])	—	1.407
s-trans-Butadiene		1.467[e])		1.473

Table 25 (continued)

Compound	Bond	Experimental Bond length	cal[d] CNDO/2	cal[e] MINDO/2
Toluene	C—Me	1.52[b]	—	1.509
Diacetylene		1.376[b]		1.386
i-Butene		1.54[b]		1.51
trans-2-butene		1.52[b]		1.50
cis-2-Butene		—		1.50
Ethylene	C=C	1.338[b]	1.320	1.337
s-trans-Butadiene		1.343[c]	—	1.347
iso-Butene		1.34[b]	—	1.347
trans-2-Butene		1.339[b]	—	1.347
cis-2-Butene		—		1.347
Allene		(1.308,1.311)[b]	—	1.309
Formaldehyde	C=O	1.210[a]	1.251	—
Acetylene		1.206[b]	1.198	1.206
Diacetylene		1.205[b]	—	1.205
Hydrogen cyanide	C≡N	1.156[a]	1.180	—

[a] Hertzberg, G., Ref. [42].

[b] Sutton, L. E.: Tables of Interatomic Distances. London: The Chemical Society 1958 and 1965.

[c] Haugen, W., Traetteberg, M.: Acta Chim. Scand. 20, 1726 (1966).

[d] Segal, G. E.: J. Chem. Phys. 47, 1876 (1967).

[e] Dewar, M. J. S., Haselbach, E., Ref. [35].

method, although still using the virtual orbital approximation, provides excellent agreement between the calculated and observed UV transitions in substituted conjugated hydrocarbons and heterocycles (Table 27). Their inprovement seems to arise from two essential modifications. One is the introduction of a limited configuration interaction and the other is the use of larger values of β for σ than for π bonds. This latter approximation has the effect of lowering the σ occupied orbitals and raising the σ unoccupied orbitals, thus shifting the transitions involving σ orbitals to larger values. In this way the intermingling of σ and π orbitals, observed in previous calculations, is avoided.

The calculation of the spectra of sigma bonded molecules was not as thoroughly studied and so far does not seem to be very successful. In

Table 26. *Comparison of experimental and calculated force constants*

Compound	Bond	Experimental[a] Force Constant (dynes/ cm $\times 10^5$)	Cal. CNDO/2[c] Force constant (dynes/ cm $\times 10^5$)	Cal. MINDO/2[d] Force Constant (dynes/ cm $\times 10^5$)
Methane	C—H	5.0	—	5.7
Ethane		4.8	12.7	5.6
Ethylene		5.1	12.8	5.8
Acetylene		5.9	12.3	6.3
Hydrogen cyanide		5.8	13.2	—
Formaldehyde		5.3	11.7	—
Ethane	C—C	4.5	33.9	5.4
Ethane (ecl)		—	—	5.5
Ethylene		9.6	23.9	10.1
Acetylene		15.8	35.5	15.9
Propane			—	5.6
n-Butane			—	5.5
t-Butadiene			—	5.6
Cyclopropane			—	5.6
Cyclopentane			—	6.3
Cyclohexane (chair)			—	6.2
Benzene		7.6[b]	—	9.4
Toluene	C—Me	—	—	5.3
Formaldehyde	C=O	12.1	34.1	
Hydrogen cyanide	C≡N	17.9	44.0	

[a]) Herzberg, G., Ref. [42].
[b]) Crawford, Jr., B. L., Miller, F. A.: J. Chem. Phys. *17*, 249 (1949).
[c]) Table 25, ref. [d]).
[d]) Dewar, M. J. S., Haselbach, E., Ref. [35].

these cases, however, the transitions occur in the far UV and the experimental data have not been as firmly established as those involving π electrons. Nevertheless, for some selected molecules they are available and correlations have been attempted. Thus Sandorfy and Katagiri [54], who developed one of the earlier modified Pariser-Parr-Pople methods including all-valence electrons, have been able to roughly reproduce the trends of the experimental transition energies for a few paraffinic molecules.

No attempts (unsuccessful?) have been reported by Del Bene and Jaffe's method in this area. INDO methods and the related EMDZO developed by Dixon have been used only casually, but seem to have led to "reasonable" values for the lowest excitation energies in a few small molecules.

Table 27. *Selected examples of correlation between observed and calculated UV transitions (Del Bene and Jaffe).* (Other studied compounds include benzene, pyridine, 1,2-diazine, 1,3-diazine, 1,4-diazine, cyclopentadienide ion, pyrazole, imidazole, 2-pyrrole-carboxaldehyde, furfural, benzonitrile, nitrosobenzene, phenol, phenoxide ion, pyridinium ion, 1-hydroxy-pyridinium ion, 2-cyano-pyridine, 3-cyano-pyridine, 4-cyano-pyridine, 2-amino-pyridine, 3-amino-pyridine, 4-amino-pyridine.)

Compound	Symmetry	Energy above the ground state (eV)		Type
		Obs	Calc	
Cyclopentadiene	$^1B_2$	4.8	4.8	$\pi \rightarrow \pi^*$
	$^1A_1$	6.2	6.3	$\pi \rightarrow \pi^*$
	$^1B_1$	7.5	7.4	$\sigma \rightarrow \pi^*$
	$^1A_1$	7.9	7.9	$\pi \rightarrow \pi^*$
Pyrrole	$^1B_2$	5.7	5.0	$\pi \rightarrow \pi^*$
	$^1A_1$	6.5	5.4	$\pi \rightarrow \pi^*$
	$^1B_2$	7.1	7.0	$\pi \rightarrow \pi^*$
	$^1A_1$		7.0	$\pi \rightarrow \pi^*$
Furan	$^1B_2$	5.9	5.2	$\pi \rightarrow \pi^*$
	$^1A_1$	6.5	5.8	$\pi \rightarrow \pi^*$
	$^1A_1$	7.4	7.3	$\pi \rightarrow \pi^*$
	$^1B_2$		7.3	$\pi \rightarrow \pi^*$
Toluene	$^1A''$	4.6	4.6	$\pi \rightarrow \pi^*$
	$^1A'$	6.0	5.1	$\pi \rightarrow \pi^*$
	$^1A'$		6.8	$\pi \rightarrow \pi^*$
	$^1A''$		6.8	$\pi \rightarrow \pi^*$

Table 27 (continued)

Compound	Symmetry	Energy above the ground state (eV)		Type
		Obs	Calc	
Aniline	$^1B_2$	4.4	4.4	$\pi \to \pi^*$
	$^1A_1$	5.4	4.7	$\pi \to \pi^*$
	$^1A_1$	6.4	6.5	$\pi \to \pi^*$
	$^1B_2$	6.9	6.6	$\pi \to \pi^*$
Anilinium ion	$^1A''$	4.9	4.6	$\pi \to \pi^*$
	$^1A'$	6.1	5.0	$\pi \to \pi^*$
	$^1A''$		6.8	$\pi \to \pi^*$
	$^1A'$		6.8	$\pi \to \pi^*$
Pyridine N-oxide	$^1A_2$	~3.8	3.2	$n \to \pi^*$
	$^1A_1$	4.4	3.9	$\pi \to \pi^*$
	$^1B_2$		4.0	$\pi \to \pi^*$
	$^1B_2$	4.9	5.8	$\pi \to \pi^*$
	$^1A_1$	6.0	6.0	$\pi \to \pi^*$

F. Nuclear Magnetic Resonance Spectra

It has been found that a good estimate of $\sigma_A{}^{13}$, the ^{13}C chemical shift (the displacement of the resonance of a carbon nucleus from a reference position when in the presence of an applied magnetic field) is a function of the electronic environment of the atom and can be correlated with its charge density.

Karplus and Pople [55], using one-electron theory, have shown that, to a first approximation, the ^{13}C chemical shift at a carbon atom A may be calculated by the following expression:

$$\sigma^{13} C_{calc}^A = \sigma_d^A + \sigma_p^A \tag{85}$$

Here, σ_d^A is a diamagnetic term proportional to the electron density of carbon atom A and expressed as

$$\sigma_d^A = 57.85 - 8.23 \, \Delta q^A \tag{86}$$

where Δq^A, the excess charge density, is given by

$$\Delta q^A = 4 - \sum_{i}^{\substack{\text{orbitals} \\ \text{on atom A}}} P_{ii}^A \qquad (87)$$

P_{ii}^A being the charge density or occupancy of orbital i of atom A in the molecule. The second term, σ_p^A, a paramagnetic term representing a local correction for the molecular environment, involves the mixing of ground and excited electronic states. This term is extremely difficult to calculate and no exact expression has been found using many-electron theory. Karplus and Pople have represented σ_p^A as

$$\sigma_p^A = -(103.57 + 33.46\ \Delta q^A) \sum_{AB} Q_{BA} \qquad (88)$$

where

$$\sum_{AB} Q_{BA} = \tfrac{4}{3}(P_{x_A x_A} + P_{y_A y_A} + P_{z_A z_A}) - \tfrac{2}{3}(P_{y_A y_A} P_{z_A z_A} + P_{z_A z_A} + P_{x_A x_A}$$

$$- \tfrac{4}{3}(P_{x_A y_B} P_{y_A x_B}) + \tfrac{2}{3}(P_{x_A y_A} P_{y_A x_A})$$

$$- \sum_{B \neq A} \tfrac{2}{3}(P_{y_A y_B} P_{z_A z_B} + P_{z_A z_B} P_{x_A x_B} + P_{x_A x_B} P_{y_A y_B})$$

and $P_{x_A y_B}$ is the bond order between a $2\,px$ atomic orbital an atom A and a $2\,py$ atomic orbital on atom B.

As the variation in σ_d^A can be shown not to total more than 20 parts per million, the paramagnetic term σ_p^A must make the dominant contribution to the chemical shift. (This was also the conclusion reached for ^{19}F shifts [56].

Both terms, however, are dependent on the total charge density of the atom. It is not surprising, therefore, that ^{13}C shifts of atoms in conjugated molecules vary approximately linearly with the π-electron density at the atoms $(\delta^{13}C = 160\ \Delta q^\pi)$. Of the available all-valence electron methods, chemical shifts have been calculated only by the CNDO approximation.

Bloor and Breen [49] calculations (CNDO/2) of the ^{13}C shifts fo monosubstituted benzenes and oxygen and nitrogen heterocycles, show that the correlation between the theoretical and experimental shifts (Eq. 85) is not significantly better than a simpler correlation of the experimental shifts with Δ_q^{total}, the total electron density. The calculated and experimental ^{13}C shifts are given in Tables 28 to 30.

Table 28. *Calculated and observed* ^{13}C *chemical shifts[a])* *for heterocyclic compounds with one ring* [b])

Compound	Atom No.	δ (^{13}C) exp (ppm)	δ (^{13}C) cal CNDO/2 (ppm)
Benzene		0	0
	2	−21.85	−11.7
	3	+4.29	+ 3.85
	4	− 7.63	− 6.11
	2	−17.4	− 7.52
	3	−23.9	− 8.47
	4	+ 1.08	+ 0.24
	1	−33.4	−13.97
	3	−30.48	−21.9
	1	−28.87	−15.0
	6	+ 6.1	+ 7.56
	2	−39.0	−25.89
	2	−14.4	− 5.12
	3	+ 1.6	+ 9.79
	2	+ 9.4	− 1.86
	3	+19.6	+ 7.30

[a]) Parts per million relative to benzene.
[b]) Bloor, J. E., Breen, D. L., reference [49a]).

Table 29. *Calculated and observec* [13] *chemical shifts*[a]) *for quinoline and isoquinoline*[b])

Compound	Atom No.	δ (^{13}C) exp (ppm)	δ (^{13}C) cal CNDO/2 (ppm)
	2	−23.0	−11.86
	3	+ 8.0	+ 4.38
	4	− 7.5	− 2.47
	5	+ 2.0	+ 0.45
	6	+ 2.5	+ 7.8
	7	− 1.0	+10.0
	8	− 1.5	+ 1.52
	9	−21.5	−30.92
	10	+ 1.0	−18.08
	1	−24.5	−11.32
	3	−15.0	−10.54
	4	− 7.0	+55.25
	5	+ 2.2	+ 5.92
	6	− 2.0	− 0.68
	7	+ 1.2	+ 0.75
	8	+ 1.0	+18.25
	9	+ 0.03	−17.78
	10	− 7.4	−22.78

[a]) Parts per million relative to benzene.
[b]) Bloor, J. E., Breen, D. L., Ref. [49a]).

Some ^{19}F chemical shifts have been reproduced for a number of substituted fluorobenzenes by Davies [50]) (CNDO/2) using the following Prosser-Goodman expression for $\sigma(X)$, the chemical shift of a compound X relative to fluorobenzene:

$$\sigma\,(X) - \sigma\,(C_6H_5F) = \frac{\sigma_0}{\varDelta w}\,(A\,\varDelta q_F + B\,\varDelta_{PFC} + C\,\varDelta q_C) \qquad (89)$$

Here $\varDelta q_F = q_F\,(X) - q_F\,(C_6H_5F)$ is the π-electron charge density on the fluorine atom in X relative to fluorobenzene; $\varDelta q_{FC}$ is the corresponding difference in π-electron bond order for the FC bond; $\varDelta q_C$ is the corresponding difference in π charge density on the neighboring carbon atom, A, B and C are parameters and $\varDelta w$ is the average energy (parameter). The results of this work are shown in Table 31.

Table 30. *Calculated and observed* [13]*C chemical shifts*[a]) *for monosubstituted benzenes*[b])

Substituent	Atom No.	δ][13]C) exp (ppm)	δ (^{13}C) cal CNDO/2 (ppm)
H	1	0	0
CH$_3$	1	$-$ 9.1	-18.2
	2	$-$ 0.3	$+$ 1.0
	4	$+$ 2.8	$+$ 0.8
F	1	-35.1	-34.5
	2	$+14.3$	$+$ 6.4
	3	$-$ 0.9	$+$ 2.8
	4	$+$ 4.4	$+$ 1.4
OH (planar)	1	-26.9	-32.0
	2	$+12.3$	$+$ 6.7
	3	$-$ 1.7	$-$ 8.8
	4	$+$ 7.3	$+$ 2.1
NH$_2$ (pyrimidal)	1	-19.2	-26.0
	2	$+12.4$	$+$ 5.3
	3	$-$ 1.3	$-$ 2.4
	4	$+$ 9.5	$+$ 2.3
CHO	1	$-$ 9.0	-13.5
	2	$-$ 1.2	$-$ 1.9
	3	$-$ 1.2	0
	4		$-$ 1.3
NO$_2$	1	-19.6	-15.3
	2	$+$ 5.3	$+$ 0.7
	3	$-$ 0.8	$-$ 1.5
	4	$-$ 6.0	$-$ 1.5

[a]) Parts per million relative to benzene.
[b]) Bloor, J. E., Breen, D. L., Ref. [49b]).

Table 31. *Calculated and observed* ^{19}F *chemical shifts* [a, b]

Compounds	Atom No.	δ (^{19}F) exp (ppm)	δ (^{19}F) cal CNDO/2 (ppm)
o-$C_6H_4F_2$	—	22.5	22.2
m-$C_6H_4F_2$	—	− 3.1	− 4.3
p-$C_6H_4F_2$	—	6.8	(6.8)
1,2,3-$C_6H_3F_3$	1	19.3	17.5
	2	46.1	46.0
1.2.4-$C_6H_3F_3$	1	27.5	30.7
	2	17.6	17.4
	4	2.8	1.7
1,2,3,4,5-C_6HF_5	1	23.4	19.8
	2	46	48.9
	3	37.6	36.0
C_6F_6	—	46.4	44.3
o-$C_6H_4FNO_2$	—	5.6	6.4
m-$C_6H_4FNO_2$	—	− 3.4	2.1
p-$C_6H_4FNO_2$	—	− 9.3	− 0.6

[a]) Davies, D. W., Ref. [51].
[b]) Relative to C_6H_5F

G. Electron Spin Resonance Spectra

The electron spin resonance (E.S.R.) spectra of a paramagnetic organic molecule, e.g. free radical, radical cation or radical anion, is directly related to its unpaired electron distribution (spin density). In the region of a magnetic nucleus the hyperfine interaction between the magnetic moments of the nucleus and the electron is a function of the spin density. It has been shown that, for an atom N, a direct correlation exists between a_N, its observed hyperfine coupling constant, and (p^a-p^β), the unpaired electron population of its atomic orbitals [d]).

At the CNDO/2 level Davies [50]) has calculated spin densities for some fluorinated nitrobenzenes and shown that they follow the general pattern of hyperfine coupling constants.

[d]) The isotropic part of the hyperfine coupling constant is related to the unpaired s electron; the anisotropic part of it is related to the unpaired electronic population of p orbitals.

Pople *et al* [27]) developed the INDO method specifically to account for hyperfine coupling constants in terms of spin densities. Pople, Beveridge and Dobosh have calculated the spin densities of a large number of compounds and found a good agreement with the experimental isotropic hyperfine coupling constants of ^1H, ^{13}C, ^{14}N, ^{17}O, and ^{19}F (Tables 32 to 35).

Beveridge and Miller [51]) in an INDO study, have calculated the vibronic effects in substituted methyl redicals and have satisfactorily reproduced the trends of isotope effects on isotropic coupling constants (Table 37).

Table 32. *Selected[a]) examples of correlation between experimental and calculated isotropic hyperfine coupling constants for* ^1H

Radical	Atom or group	a_N, G Calcd	Exptl
Methyl		−22.4	(−)23.04
Fluoromethyl		− 7.8	(−)21.10
Difluoromethyl		21.9	(+)22.20
Ethyl	CH_2	−20.4	(−)22.38
	CH_2	27.6	(+)26.87
Vinyl	α	17.1	(+)13.40
	β_1	55.1	(+)65.00
	β_2	21.2	(+)37.00
Formyl		74.9	(+)137.00
Ethynyl		32,7	(+)16.10
Allyl	1	−14.6	(−)13.93
	1′	−14.9	(−)14.83
	2	6.9	(+) 4.06
Phenyl	2	18.7	(+)19.50
	3	6.1	(+) 6.50
	4	3.9	
Cyclopentadienyl		− 4.8	(−) 5.60
Tropyl		− 3.2	(−) 3.95
Benzyl	−CH_2	−17.0	(−)16.35
	2	− 6.4	(−) 5.14
	3	3.6	(+) 1.75
	4	− 5.6	(−) 6.14
Phenoxy	2	− 4.1	(+) 6.60
	3	2.2	(+) 1.96
	4	− 3.4	(−)10.40

Table 32 (continued)

Radical	Atom or group	a_N, G Calcd	Exptl.
Cyclohexadienyl	CH_2	97.6	(+)47.71
	2	−11.1	(+) 8.99
	3	5.1	(+) 2.65
	4	− 9.8	(−)13.04
Perinaphthenyl	1	− 7.5	(−) 7.30
	2	4.3	(+) 2.80
Benzene⁻		− 3.6	(−) 3.75
Cyclooctatetraene⁻		− 2.6	(−) 3.21
trans-Butadiene⁻	1	− 9.8	(−) 7.62
	1′	−10.3	(−) 7.62
	2	− 0.8	(−) 2.79
Naphthalene⁻	1	− 5.3	(−) 4.90
	2	− 0.9	(−) 1.83
Anthracene⁻	1	− 2.7	(−) 2.74
	2	− 0.6	(−) 1.51
	9	− 6.8	(−) 5.34
Anthracene⁺	1	− 2.9	(−) 3.00
	2	− 0.6	(−) 1.38
	9	− 6.6	(−) 6.49
Phenanthrene⁻	1	− 4.6	(−) 3.60
	2	1.2	(+) 0.72
	3	− 3.8	(−) 2.88
	4	0.6	(+) 0.32
	9	− 5.0	(−) 4.32
Pyrene⁻	1	− 5.5	(−) 4.75
	2	2.5	(+) 1.09
	4	− 1.9	(−) 2.08
Stilbene⁻	1	− 3.7	(−) 1.90
	2	2.0	(+) 0.86
	3	− 3.9	(−) 3.80
	4	1.9	(+) 0.32
	5	− 3.4	(−) 2.95
	7	− 5.2	(−) 4.36
Biphenylene⁻	1	0.2	(+) 0.21
	2	− 2.1	(−) 2.86
Azulene⁻	1	0	(+) 0.27
	2	− 3.0	(−) 3.95
	4	− 7.0	(−) 6.22
	5	3.9	(+) 1.34
	6	− 9.4	(−) 8.82
Fluoranthene⁻	1	− 4.4	(−) 3.90
	2	2.2	(+) 1.30
	3	− 6.4	(−) 5.20
	7	0.2	
	8	− 0.9	

Table 32 (continued)

Radical	Atom or group	a_N, G Calcd	Exptl
Benzonitrile⁻	2	− 3.3	(−) 3.63
	3	1.1	(+) 0.30
	4	− 8.0	(−) 8.42
Phthalonitrile⁻	3	1.5	(+) 0.33
	4	− 4.0	(−) 4.24
Isophthalonitrile⁻	2	1.5	(+) 0.08
	4	− 7.6	(−) 8.29
	5	2.6	(+) 1.44
Terephthalonitrile⁻		− 1.0	(−) 1.59
1,2,4,5-Tetracyanobenzene⁻		2.2	(+) 1.11
p-Nitrobenzonitrile⁻	2	1.8	(+) 0.76
	3	− 3.5	(−) 3.12
Nitrobenzene⁻	2	− 3.6	(−) 3.39
	3	1.9	(+) 1.09
	4	− 3.8	(−) 3.97
m-Dinitrobenzene⁻	2	0.4	(+) 3.11
	4	− 7.8	(−) 4.19
	5	3.2	(+) 1.08
p-Dinitrobenzene⁻		− 1.0	(−) 1.12
m-Fluoronitrobenzene⁻	2	− 3.7	(−) 3.30
	4	− 3.7	(−) 3.30
	5	1.8	(+) 1.10
	6	− 3.4	(−) 3.00
p-Fluoronitrobenzene⁻	2	− 3.8	(−) 3.56
	3	2.2	(+) 1.16
3,5-Difluoronitrobenzene⁻	2	− 3.5	(−) 3.26
	4	− 3.6	(−) 3.98
o-Benzosemiquinone⁻	3	− 1.9	(−) 3.65
	4	0.2	(+) 0.95
p-Benzosemiquinone⁻		− 0.9	(−) 2.37
2,5-Dioxo-1,4-semiquinone²⁻		2.4	(+) 0.79
1,4-Naphthosemiquinone⁻	2	− 1.0	(−) 3.23
	5	0.6	(+) 0.65
	6	− 0.1	(−) 0.51

a) Other compounds whose hyperfine coupling constants have been correlated include 9,10-anthrasemiquinone⁻, pyrazine⁻, 1,5-diazonaphthalene⁻, pyridazine⁻, 5-tetrazine⁻, N,N-dihydropyrazine, phthalazine, quinoxaline⁻, dihydroquinoxaline⁺, phenazine⁻, 1,4,5,8-tetraazoanthracene⁻, p-nitrobenzaldehyde⁻, p-cyanobenzaldehyde⁻, and 4-cyanopyridine⁻.

Table 33. *Selected*[a]) *examples of correlation between experimental and calculated isotropic hyperfine coupling constants for* ^{13}C

Radical	Atom or group	a_N, G Calcd	Exptl
Methyl		45.0	(+) 38.34
Fluoromethyl		92.7	(+) 54.80
Difluoromethyl		145.1	(+)148.80
Trifluoromethyl		184.6	(+)271.60
Ethyl	CH_3	− 12.4	(−) 13.57
	CH_2	39.9	(+) 39.07
Vinyl	α	178.0	(+)107.57
	β	− 14.5	(−) 8.55
Ethynyl	1	− 2.5	
	2	342.8	
Allyl	1	23.0	
	2	− 16.6	
Phenyl	1	151.3	
	2	− 4.8	
	3	10.7	
	4	− 2.6	
Cyclopentadienyl		4.1	
Tropyl		3.5	
Benzyl	1	− 12.3	
	2	11.7	
	3	− 8.5	
	4	10.5	
	CH_2	32.6	
Phenoxy	1	− 10.7	
	2	7.0	
	3	− 5.5	
	4	6.3	
Cyclohexadienyl	2	17.9	
	3	− 13.7	
	4	17.8	
	CH_2	− 17.6	
Perinaphthenyl	1	13.9	
	2	− 10.3	
	4	− 9.3	
	13	6.7	
Benzene⁻		4.0	(+) 2.80
Cyclooctatetraene⁻		3.0	(+) 1.28
trans-Butadiene⁻	1	18.6	
	2	− 1.2	
Naphthalene⁻	α	9.3	(+) 7.10
	β	− 0.3	(−) 1.20
	9	− 4.3	
Anthracene⁻	1	4.6	3.57
	2	0	− 0.25
	9	12.4	8.70
	11	− 3.4	− 4.59

Table 33 (continued)

Radical	Atom or group	a_N, G Calcd	Exptl
Anthracene$^+$	2	0.2	(+) 0.37
	9	11.8	8.48
	11	— 3.3	(—) 4.50
Phenanthrene$^-$	1	8.2	
	2	— 5.7	
	3	— 6.9	
	4	— 2.2	
	9	7.5	
	11	— 3.8	
	12	2.1	
Pyrene$^-$	1	9.9	
	2	— 7.1	
	4	2.9	
Stilbene$^-$	1	6.2	
	2	— 5.2	
	3	7.4	
	4	— 4.9	
	5	5.8	
	6	— 3.2	
	7	7.4	
Biphenylene$^-$	1	— 3.0	
	2	3.0	
	10	5.2	
Azulene$^-$	1	— 1.8	
	2	4.9	
	9	1.3	
	4	11.7	
	5	— 10.2	
	6	16.9	
Fluoranthene$^-$	1	7.5	
	2	— 6.4	
	3	12.0	
	7	— 1.2	
	8	1.3	
	11	— 7.0	
	12	1.6	
	13	— 0.4	
	14	2.4	
Benzonitrile$^-$	1	8.4	
	2	3.6	
	3	— 5.2	
	4	14.0	
	CN	— 6.6	(—) 6.12
Phthalonitrile$^-$	1	8.5	
	3	— 6.1	
	4	6.0	
	CN	— 6.4	

Table 33 (continued)

Radical	Atom or group	a_N, G	
		Calcd	Exptl
Isophthalonitrile⁻	1	4.9	
	2	− 5.8	
	4	12.3	
	5	− 9.1	
	CN	− 4.3	
Terephthalonitrile⁻	CN	− 6.7	(−) 7.83
	1	9.7	8.81
	2	− 0.7	(−) 1.98
1,2,4,5-Tetracyanobenzene⁻	1	7.2	
	3	− 7.3	
	CN	− 5.3	
p-Nitrobenzonitrile⁻	1	7.5	
	2	− 5.2	
	3	5.5	
	4	− 2.3	
	CN	− 4.5	
Nitrobenzene⁻	1	− 5.2	
	2	6.1	
	3	− 5.2	
	4	7.1	
m-Dinitrobenzene⁻	1	0.3	
	2	− 2.4	
	4	13.2	
	5	− 9.4	
p-Dinitrobenzene⁻	1	6.1	
	2	0.1	
o-Benzosemiquinone⁻	1	− 6.6	
	3	3.2	
	4	− 1.1	
p-Benzosemiquinone⁻	1	− 6.9	(−) 0.59
	2	1.0	(+) 0.40
2,5-Dioxo-1,4-benzosemiquinone⁻	1	3.1	
	3	− 7.9	
1,4-Naphthosemiquinone⁻	1	− 8.3	
	2	1.3	
	5	− 1.5	
	6	0.2	
	9	1.4	

a) Other compounds whose hyperfine coupling constants have been
correlated include 9,10-anthrasemiquinone⁻, pyrazine⁻, 1,5-diazo-
naphthalene⁻, pyridazine⁻, 5-tetrazine⁻, N,N-dihydropyrazine⁺,
phthalazine, quinoxaline⁻, dihydroquinoxaline⁺, phenazine⁻,
1,4,5,8-tetraazoanthracene⁻, p-nitrobenzaldehyde⁻, p-cyanobenz-
aldehyde⁻, and 4-cyanopyridine⁻.

Table 34. *Observed and calculated isotropic hyperfine coupling constants for* ^{14}N

Radical	Group	a_N, G	
		Calcd	Exptl
Benzonitrile⁻		2.4	(+) 2.15
Phthalonitrile⁻		1.9	(+) 1.80
Isophthalonitrile⁻		1.3	(+) 1.02
Terephthalonitrile⁻		2.0	(+) 1.81
1,2,4,5-Tetracyanobenzene⁻		1.4	(+) 1.15
p-Nitrobenzonitrile⁻	CN	1.1	(+) 0.76
	NO₂	4.7	(+) 7.15
Nitrobenzene⁻		7.1	(+)10.32
m-Dinitrobenzene⁻		0.5	(+) 4.68
p-Dinitrobenzene⁻		−0.0	(−) 1.74
m-Fluoronitrobenzene⁻		6.6	(+)12.60
p-Fluoronitrobenzene⁻		7.1	(+) 9.95
3,5-Difluoronitrobenzene⁻		6.1	(+) 8.09
Pyrazine⁻		8.3	(+) 7.21
N,N-Dihydropyrazine⁺		7.8	(+) 7.60
Pyridazine⁻		7.7	(+) 5.90
s-Tetrazine⁻		5.8	(+) 5.28
1,5-Diazanaphthalene⁻		5.9	(+) 3.37
Phthalazine⁻		0.3	(+) 0.88
Quinoxaline⁻		7.3	(+) 5.64
Dihydroquinoxaline⁺		7.7	(+) 6.65
Phenazine⁻		7.2	(+) 5.14
1,4,5,8-Tetraazaanthracene⁻		3.3	(+) 2.41
p-Dicyanotetrazine⁻	Ring	5.9	(+) 5.88
	CN	−0.9	(−) 0.16
p-Nitrobenzaldehyde⁻		−0.5	(+) 5.83
p-Cyanobenzaldehyde⁻		1.0	(+) 1.40
4-Cyanopyridine⁻	Ring	8.3	(+) 5.67
	CN	2.7	(+) 2.33

Table 35. *Observed and calculated isotropic hyperfine coupling constants for* ^{17}O

Radical	a_N, G	
	Calcd	Exptl
p-Benzosemiquinone⁻	−8.7	(−)9.53
1,4-Naphthosemiquinone⁻	−9.3	(−)8.58
9,10-Anthrasemiquinone⁻	−9.9	(−)7.53
2,5-Dioxo-1,4-semiquinone³⁻	−3.6	(−)4.57
Nitrobenzene⁻	−4.3	(−)8.84

Table 36. *Observed and calculated isotropic hyperfine coupling constants for* ^{12}F

Radical	Atom	a_N, G	
		Calcd	Exptl
Fluoromethyl		71.3	(+) 64.30
Difluoromethyl		87.1	(+) 84.20
Trifluoromethyl		159.5	(+)142.40
Monofluoroacetamide		34.4	54.60
Difluoroacetamide	1′	31.5	75.00
	1	39.0	75.00
m-Fluoronitrobenzene⁻		−4.0	(−) 3.70
p-Fluoronitrobenzene⁻		6.3	(+) 3.41
3,5-Difluoronitrobenzene⁻		−3.8	(−) 2.73

Table 37. *Calculated and observed isotropic hyperfine coupling constants a_N for methyl radical and isotopically substituted derivatives*

Radical	Temp. (deg. K)	a_H (gauss)		a_D (gauss)		a_C (gauss)	
		Calc.	Obs.	Calc.	Obs.	Calc.	Obs.
$^{12}CH_3$	96	−22.96	(−)23.038 ± 0.01	—	—	—	—
$^{13}CH_3$	96	−22.97	(−)23.04 ± 0.01	—	—	38.06	(+)38.34 ± 0.01
$^{12}CH_2D$	85	−23.06	(−)23.10 ± 0.01	−3.540	(−)3.531 ± 0.01	—	—
$^{13}CH_2D$	85	−23.08	(−)23.10 ± 0.025	−3.543	(−)3.54 ± 0.025	37.59	(+)37.82 ± 0.025
$^{12}CD_2H$	85	−23.18	(−)23.21 ± 0.01	−3.558	(−)3.552 ± 0.01	—	—
$^{13}CD_2H$	85	−23.20	(−)23.19 ± 0.01	−3.561	(−)3.55 ± 0.01	37.04	(+)37.06 ± 0.01
$^{12}CD_3$	96	—	—	−3.579	(−)3.576 ± 0.01	—	—
$^{13}CD_3$	96	—	—	−3.581	(−)3.578 ± 0.01	36.53	(+)35.98 ± 0.01

V. Final Remarks

In the preceding pages, we have reviewed some of the most important all-valence electron methods proposed for the S.C.F. calculation of properties of large organic molecules.

The last five years have seen the birth of such methods and an incredibly fast development of a number of more efficient variants designed to give better agreement with specific properties. So far, however, no method seems to be general enough to overshadow all the others, although some of the newly developed ones seem to come closer to this ultimate goal. Below we give what we consider to be, at the present time, the most useful methods for various specific purposes.

Ionization potentials	MINDO/2
Heats of formation	MINDO/2, PNDO
Dipole moments	CNDO/2
Bond distances	CNDO/2, MINDO/2
Bond angles	CNDO/2
Force constants	CNDO/2, MINDO/2
Ultraviolet	CNDO/2 (Del Bene and Jaffe)
Nuclear Magnetic Resonance	CNDO/2
Electron Spin Resonance	INDO

The MINDO/2 method seems to be particularly attractive but, as yet, very little information is available on its applicability to some properties such as spectra.

The trend is undoubtedly in favor of the development of an "all-purpose" method, but the means by which this can be achieved are still debatable. Some authors believe that the direction to follow involves the development of an NDDO method. Such a procedure, however, would require the calculation of a much larger number of integrals and therefore would jeopardize the possibility of application to large organic molecules of „chemical interest."

It is the opinion of the present authors that such calculation would not improve the agreement with experimental properties because it would not introduce any fundamentally new feature which might correct for the inadequacies of the present ones. As a matter of fact, the neglect of two-center integrals involving one-center differential overlap seems to be a reasonable hypothesis as shown by the success of the M(INDO) methods.

On the other hand, researchers have usually confined themselves to trying to find the best approximation for molecular integrals but generally overlooked the possibility that atomic orbitals in molecules might differ widely from those in the isolated atoms.

It is thus one of the common features of all methods described so far (see however Wiberg's CNDO [30]) to) determine atomic parameters from the atomic spectra. A close analysis of the shortcomings of the existing methods shows, however, that this might not be very appropriate and a better assessment of atomic parameters in molecules might offer a successful new route.

Acknowledgment. We express our sincere thanks to the National Science Foundation for financial support of this work through Grant No. GP-8513.

VI. References

1) Lennard-Jones, J. E.: Trans. Faraday Soc. *25*, 668 1929).
2) Mulliken, R. S.: J. Chem. Phys. *3*, 375 (1935).
3) Hückel, E.: Z. Physik *70*, 204 (1931); *72*, 310 (1931); *76*, 628 (1932).
4) Goeppert-Mayer, M., Sklar, A. L.: J. Chem. Phys. *6*, 645 (1938).
5) Wheland, G. W., Mann, D. E.: J. Chem. Phys. *17*, 264 (1949).
6) Streitwieser, A., Jr.: J. Am. Chem. Soc. *82*, 4123 (1960).
7) Pariser, R., Parr, R. G.: J. Chem. Phys. *21*, 466, 767 (1953).
8) Pople, J. A.: Trans. Faraday. Soc. *49*, 1375 (1953).
9) Roothaan, C. C. J.: Rev. Mod. Phys. *23, 69* (1951).
10) For a review of these methods see: Klopman, G.: Tetrahedron *19*, 111 (1963).
11) Brown, R. D.: J. Chem. Soc. 2615 (1953).
12) Dewar, M. J. S., Pettit, R.: J. Chem. Soc. 1617 (1954).
13) Hall, G. G., Lennard-Jones, J. E.: Proc. Roy. Soc. *A 202*, 155 (1950); *A 205*, 357 (1951).
14) Franklin, J. L.: J. Chem. Phys. *22*, 1304 (1954).
15) Sandorfy, C., Daudel, R.: Compt. Rend. *238*, 93 (1954).
16) — Can. J. Chem. *33*, 1337 (1955).
17) Yoshimuzi, H.: Trans. Faraday Soc. *53*, 125 (1957).
18) Fukui, K., Kato, H., Yonezawa, T.: Bull. Chem. Soc. Japan *33*, 1701 (1960); *34*, 442, 1111 (1961).
19) Klopman, G.: Helv. Chim. Acta *45*, 711 (1962); *46*, 1967 (1963).
20) Wolfsberg, M., Helmholtz, L.: J. Chem. Phys. *20*, 837 (1952).
21) Mulliken, R. S.: J. Phys. Chem. *56*, 792 (1952).
22) Hoffmann, R.: J. Chem. Phys. *39*, 1397 (1963).
23) Pohl, H. A., Appel, R., Appel, K.: J. Chem. Phys. *41*, 3385 (1964).
24) Klopman, G.: J. Am. Chem. Soc. *86*, 1463, 4550 (1964); *87*, 3300 (1965).
25) Pople, J. A., Santry, D. P., Segal, G. A.: J. Chem. Phys. *43*, 5129 (1965).
26) — Segal, G. A.: J. Chem. Phys. *44*, 3289 (1966).
27) — Beveridge, D. L., Dobosh, P. A.: J. Chem. Phys. *47*, 2026 (1967).
28) Dewar, M. J. S., Klopman, G.: J. Am. Chem. Soc. *89*, 3089 (1967).
29) a) Fisher, H., Kollmar, H.: Theor. Chim. Acta *13*, 213 (1969);
   b) Clark, D. T.: Theor. Chim. Acta *10*, 111 (1968);
   c) Lo, D. H., Whitehead, M. A.: Can. J. Chem. *46*, 2027 (1968);
   d) Jaffe, H. H.: Acc. Chem. Res. *2*, 136 (1969).
30) Wiberg, K.: J. Am. Chem. Soc. *90*, 59 (1968).
31) Del Bene, J., Jaffe, H. H.: J. Chem. Phys. *48*, 1807, 4050 (1968); *49*, 1221 (1968).
32) a) (CNDO level) Yonezawa, T., Yamaguchi, K., Kato, H.: Bull. Chem. Soc. Japan *40*, 536 (1967);
   b) (INDO level) Kato, H., Konichi, H., Yamale, H., Yonezawa, T.: Bull. Chem. Soc. Japan *40*, 2761 (1967).

33) Dixon, R. N.: Mol. Phys. *12*, 83 (1967).

34) Baird, N. C., Dewar, M. J. S.: J. Chem. Phys. *50*, 1262 (1969).

35) Dewar, M. J. S., Haselbach, E.: Private Communication, J. Am. Chem. Soc. Feb. 1970.

36) Clementi, E., IBM: J. Res. Develop. *9*, 2 (1965).

37) Parr, R. G.: J. Chem. Phys. *20*, 1499 (1952).

38) Dewar, M. J. S., Wulfman, C. E.: J. Chem. Phys. *29*, 158 (1958).

39) Nishimoto, K., Nataga, N.: Z. Physik. Chem. *12*, 335 (1957).

40) Ohno, K.: Theoret. Chim. Acta *2*, 219 (1964).

41) Al Joboury, M. I., Turner, D. W.: J. Chem. Soc. 5141 (1963); 4434 (1964); 616 (1965).

42) Herzberg, G.: Spectra of Diatomic Molecules. Van Nostrand 1967.

43) Klopman, G.: unpublished results.

44) — J. Am. Chem. Soc. *91*, 89 (1969). — Olah, G. A., Klopman, G., Schlosberg, R. H.: J. Am. Chem. Soc. *91*, 3261 (1968).

45) Wiberg, K. B.: Tetrahedron *24*, 1083 (1968).

46) Pople, J. A., Segal, G. A.: J. Chem. Phys. *43*, 5136 (1965).

47) — Gordon, M.: J. Am. Chem. Soc. *89*, 4253 (1967).

48) Segal, G. A., Klein, M. L.: J. Chem. Phys. *47*, 4236 (1967).

49) a) Bloor, G. E., Breen, D. L.: J. Am. Chem. Soc. *89*, 6835 (1967);
    b) Bloor, G. D., Breen, D. L.: J. Chem. Phys. *72*, 716 (1968).

50) Davies, D. W.: Mol. Phys. *13*, 465 (1967).

51) Kroto, H. W., Santry, D. P.: J. Chem. Phys. *47*, 792 (1967).

52) — — J. Chem. Phys. *47*, 2736 (1967).

53) Kuznetsof, P. M., Shriver, D. F.: J. Am. Chem. Soc. *90*, 1683 (1968).

54) Kataziri, S., Sandorfy, C.: Theor. Chim. Acta *4*, 203 (1966).

55) Karplus, M., Pople, J. A.: J. Chem. Phys. *38*, 7803 (1963).

56) Saika, A., Slichter, C. P.: J. Chem. Phys. *22*, 76 (1954).

57) Beveridge, D. L., Muller, K.: Mol. Phys. *14*, 401 (1968).

58) Sustman, R., Williams, J. E., Dewar, M. J. S., Allen, L. C., Schleyer, P. von R.: J. Am. Chem. Soc. *91*, 5350 (1969).

59) Bodor, N., Dewar, M. J. S., Worley, S. D.: J. Am. Chem. Soc. *92*, 19 (1970).

60) Isaacs, N. S.: Tetrahedron *25*, 3555 (1969).

Received February 25, 1970

SPRINGER-VERLAG
BERLIN·HEIDELBERG·NEW YORK

HMO
Hückel Molecular
Orbitals

Von **E. Heilbronner** und **P. A. Straub**
Laboratorium für Organ. Chemie der
Eidgenössischen Technischen Hochschule
Zürich

816 Seiten. 1966
Loseblattheftung
DM 72,—
US $ 24.00

Die Tabellen enthalten die Eigenwerte, Linearkombinationen, Ladungs- und Bindungsordnungen sowie die Polarisierbarkeit einer Auswahl von π-Elektronensystem-Modellen, berechnet nach dem Hückelschen Näherungsverfahren.

Sie liefern diejenige Information, die für eine Verwendung der HMOs innerhalb einfacher Störungsrechnungen notwendig ist und sollen dem theoretisch interessierten Chemiker einen Grundstock von HMOs zugänglich machen, mit dessen Hilfe ein großer Teil der in der Praxis und während des Studiums auftretenden Probleme gelöst werden kann.

Ein Maschinenprogramm zur selbständigen Berechnung komplizierterer Beispiele ist beigefügt.

Titel-Nr. 7725

SPRINGER-VERLAG
BERLIN·HEIDELBERG·NEW YORK

THEORETICA CHIMICA ACTA

edenda curat: **Hermann Hartmann**

adiuvantibus: C. J. Ballhausen, København;
R. D. Brown, Clayton; E. Heilbronner, Basel;
J. A. A. Ketelaar, Amsterdam; M. Kotani, Tokyo;
J. Koutecký, Praha; J. W. Linnett, Cambridge;
E. E. Nikitin, Moskwa; R. G. Pearson, Evanston;
B. Pullman, Paris; K. Ruedenberg, Ames;
C. Sandorfy, Montreal; M. Simonetta, Milano;
O. Sinanoglu, New Haven

"Theoretica Chimica Acta" will publish papers dealing with the relationship of chemical and physical phenomena to the deductions made from valence and electronic theories. First consideration will be given to those which are primarily of chemical interest.

1970: 3 volumes (Vol. Nos. 16—18)
Price per volume DM 96,—; US $ 26.40
plus postage

SPRINGER-VERLAG
BERLIN·HEIDELBERG·NEW YORK

STRUCTURE AND BONDING

Editors: P. Hemmerich, Konstanz;
C. K. Jørgensen, Genève; J. B. Neilands,
Berkeley; Sir Ronald S. Nyholm, London;
D. Reinen, Bonn; R. J. P. Williams, Oxford

Vol. 7

With 45 figures
III, 154 pages
1970
Soft cover DM 38,—
US $ 10.50

■ **Prospectus
on request**

The Spectra of Ferric Haems and Haemo-
proteins. By Dr. D. W. Smith, Chemistry
Dept., The University of Sheffield, and Prof.
R. J. P. Williams, Inorganic Chemistry
Laboratory, Oxford

The Absolute Configuration of Transition
Metal Complexes. By Dr. R. D. Gillard and
Dr. P. R. Mitchell, Inorganic Chemistry
Laboratory, The University, Canterbury, Kent

The Application of Nuclear Quadrupole
Resonance Spectroscopy to the Study of
Transition Metal Compounds. By Dr. W. van
Bronswyk, William Ramsey and Ralph Forster
Laboratories, University College, Gower
Street, London W.C. 1

Kationenverteilung zweiwertiger $3d^n$-Ionen
in oxidischen Spinell-, Granat- und anderen
Strukturen. Von Dr. D. Reinen, Anorganisch-
Chemisches Institut der Universität Bonn

NMR

Basic Principles and Progress
Grundlagen und Fortschritte
Editors: P. Diehl, E. Fluck, R. Kosfeld

Vol. 1

NMR Studies of Molecules
Oriented in the Nematic Phase
of Liquid Crystals
By Professor Dr. P. Diehl and Dr. C. L. Khetrapal

The Use of Symmetry in Nuclear
Magnetic Resonance
By Dr. R. G. Jones

With 53 fig.
V, 174 pp. 1969
Cloth DM 39,—
US $ 10.80

Part one of this volume contains an introduction to the
NMR spectroscopy of molecules dissolved in the nematic
phase of liquid crystals. This new type of spectroscopy
allows the determination of direct spin-spin coupling constants. In the second part recognition of molecular symmetry and its relation to group theory are shown to be
essential components in the understanding of NMR spectra
and their analysis.

Vol. 2

NMR-Untersuchungen
an Komplexverbindungen
Von Dozent Dr. H. J. Keller

Mit 22 Abb.
III, 88 S. 1970
Geb. DM 32,—
US $ 8.80

Der Band gibt einen Überblick über die Anwendungs-
möglichkeiten der NMR-Methode auf dia- bzw. paramagnetische Komplexe. Der Verfasser geht auf die große
Bedeutung der NMR-Spektroskopie bei der Untersuchung
schnell ablaufender Reaktionen ein.

SPRINGER-VERLAG
BERLIN · HEIDELBERG · NEW YORK

SPRINGER-VERLAG
BERLIN·HEIDELBERG·NEW YORK

Scheffler/Stegmann
Elektronenspinresonanz

Grundlagen und Anwendung in der organischen Chemie

Von Dr. Klaus Scheffler und Dr. Hartmut B. Stegmann

Mit 145 Abbildungen
VIII, 506 Seiten. 1970
DM 120,–; US $ 33.00

(Organische Chemie
in Einzeldarstellungen,
Band 12)

Die Elektronenspinresonanz kann Aufschlüsse über paramagnetische organische Verbindungen liefern. Die ersten Abschnitte behandeln die theoretischen Grundlagen und die physikalischen Probleme, die als Voraussetzungen zum Verständnis der Meßergebnisse unerläßlich sind. Der zweite Teil bietet aus der Fülle der Resultate eine kritische Auswahl. Sie soll zur praktischen Ergänzung der vorstehenden Kapitel und als experimentelles Vergleichsmaterial dienen. Dergestalt vermittelt das Buch jene Kenntnisse, die zum Studium stabiler organischer Radikale und paramagnetischer Reaktions-Zwischenstufen notwendig sind.

Inhaltsübersicht:

SPRINGER-VERLAG
BERLIN·HEIDELBERG·NEW YORK

Fortschritte der chemischen Forschung
Topics in Current Chemistry

Herausgeber: A. Davison, M. J. S. Dewar, K. Hafner, E. Heilbronner, U. Hofmann, K. Niedenzu, Kl. Schäfer, G. Wittig

Schriftleitung: F. Boschke

Band 9, Heft 3

Mit 20 Abbildungen
161 Seiten. 1968
Geheftet DM 38,—
US $ 10.50

Mehrelektronen Modelle

H. Preuß, Gegenwärtige Möglichkeiten wellenmechanischer Absolutrechnungen an Molekülen und Atomsystemen. — M. Klessinger, Mehrelektronenmodelle in der organischen Chemie. — J. Hinze, Elektronegativität der Valenzzustände.

Band 10, Heft 1

Mit 49 Abbildungen
205 Seiten. 1968
Geheftet DM 58,—
US $ 16.00

Spektren und Molekülbau

R. Zahradnik, Elektronenspektren konjugierter Verbindungen im ultravioletten und sichtbaren Bereich. — H. Dreizler, Mikrowellenspektroskopische Bestimmung von Rotationsbarrieren freier Moleküle. — H. J. Becher, Kraftkonstantenberechnungen aus den Schwingungsspektren einfacher organischer Moleküle.

Volume 15, Issue 1

With 47 figures
85 pages. 1970
Soft cover DM 34,—
US $ 9.40

Orientation and Stereoselection

K. Fukui, Theory of Orientation and Stereoselection.

FORTSCHRITTE DER CHEMISCHEN FORSCHUNG
TOPICS IN CURRENT CHEMISTRY

Herausgeber:
A. Davison · M. J. S. Dewar
K. Hafner · E. Heilbronner
U. Hofmann · K. Niedenzu
Kl. Schäfer · G. Wittig
Schriftleitung: F. Boschke

15. BAND

1970

Springer-Verlag Berlin Heidelberg GmbH

Inhalt des 15. Bandes

Mitarbeiter des 15. Bandes

Prof. *J. E. Baldwin*, Department of Chemistry, University of Oregon, Eugene, OR 97403, USA

Prof. *R. H. Fleming*, Department of Chemistry, University of Oregon, OR 97403, USA ⁻

Prof. *K. Fukui*, Department of Hydrocarbon Chemistry, Kyoto University, Kyoto, Japan

Prof. Dr. *G. Heller*, Institut für Anorganische Chemie der Freien Universität Berlin, 1000 Berlin 33, Fabeckstraße 34/36

Prof. *H. D. Johnson, II*, Department of Chemistry, The Ohio State University, Columbus, OH 43210, USA

Prof. *G. Klopman*, Department of Chemistry, Case Western Reserve University, Cleveland, OH 44106, USA

Prof. Dr. *J. M. Lehn*, Institut de Chimie, Université de Strasbourg, 1, rue Blaise Pascal, B.P. 296 / R 8, F—67 Strasbourg

Dr. *A. Meller*, Institut für Anorganische Chemie, Technische Hochschule Wien, Getreidemarkt 9, A—1060 Wien

C. D. Miller, M. A., Department of Chemistry, University of Kentucky, Lexington, KY 40506, USA

Prof. Dr. *K. Niedenzu*, Department of Chemistry, University of Kentucky, Lexington, KY 40505, USA

Dr. *B. O' Leary*, Department of Chemistry, Case Western Reserve University, Cleveland, OH 44106, USA

Prof. *S. G. Shore*, Department of Chemistry, The Ohio State University, Columbus, OH 43210, USA

Doz. Dr. *W. Tochtermann*, Institut für Organische Chemie der Universität Heidelberg, 6900 Heidelberg 1, Tiergartenstraße

In kritischen Übersichten werden in dieser Reihe Stand und Entwicklung aktueller chemischer Forschungsgebiete beschrieben. Sie wendet sich an alle Chemiker in Forschung und Industrie, die am Fortschritt ihrer Wissenschaft teilhaben wollen.

In der Regel werden nur Beiträge veröffentlicht, die ausdrücklich angefordert worden sind. Schriftleitung und Herausgeber sind aber für ergänzende Anregungen und Hinweise jederzeit dankbar. Manuskripte können in den ,,Fortschritten der chemischen Forschung'' in Deutsch oder Englisch veröffentlicht werden.

Jedes Heft der Reihe ist auch einzeln käuflich.

This series presents critical reviews of the present position and future trends in modern chemical research. It is addressed to all research and industrial chemists who wish to keep abreast of advances in their subject.

As a rule, contributions are specially commissioned. The editors and publishers will, however, always be pleased to receive suggestions and supplementary information. Papers are accepted for "Topics in Current Chemistry" in either German or English.

Single issues may be purchased separately.